THE CAMERA

From the 11th Century to the present day

THE CAMERA

From the 11th Century to the present day

By John Wade

Published by Jessop Specialist Publishing,
Jessop House,
98 Scudamore Road,
Leicester, LE3 1TZ.
First published 1990.

Design, typesetting and editorial production by
Wordpower Publishing,
PO Box 303, Welwyn, Hertfordshire, AL6 9AG.

ISBN: 0-9514392-3-5

Printed by Newnorth Print Limited,
Newnorth House,
College Street,
Kempston, Bedford, MK42 8NA.

Cover picture shows the Mammoth, the world's largest camera. Original photograph kindly supplied by Olympus UK Limited. Artwork by Steven Archibald, Sapphire Publications.

CONTENTS

The shape of cameras past - a Meagher Studio Camera from around 1890.

Introduction

THE STORY OF THE CAMERA is one that has always interested photographers. But today it interests others too - collectors, historians, investors and even antique dealers. All of whom will find something to intrigue and entertain them in this short history.

This is, it must be said, by no means the first book written on the history of the camera. But I would venture to suggest that few have attempted what you are about to read here: a concise run through camera innovation and design, starting way back in the 11th Century, to gradually following the path that led right up to today's computerised wizardry and video imaging.

For those who celebrated the 150th anniversary of photography in 1989, the idea of beginning this book in the 11th Century must seem rather strange. Personally, I found it strange that the 150th anniversary assumed 1839 - the date that Daguerre announced his process to the world - to be the start of photography. Granted that Daguerre's daguerreotypes and Talbot's calotypes were the first *viable* methods of photography, but the first photograph is can be traced back to Niépce as early as1822 (even though the first *surviving* photograph dates to a few years later). But even that didn't mark the appearance of the first camera - otherwise how could I justify saying that this book starts its story in the 11th Century? The fact is, of course, that cameras were invented long before photography, and if you want to know the how, where, when and why of that statement, you'll have to move on to chapter one.

The past decade has seen tremendous change and innovation in the history of the camera. Autofocus has arrived with a vengeance, in both simple snapshot cameras and the more sophisticated single lens reflexes. Built-in extras such as meters, flashguns and motordrives have become the norm. Automation, in fact, has taken paths that might never have been dreamt of only a few short years ago. Which is why an up-to-date book on this fascinating subject is undoubtedly due.

Not that I would want to give anyone the impression that the book you are about to read is all about the latest technology. The last chapter, it's true, is dedicated to cameras in the computer age, but essentially, this is a history book, one that brings you details of classic cameras and the processes they used, down through the ages, some of which might be surprising to today's hi-tech camera owners. For instance...

The first photograph took a day to expose.

Victorian photographers held the heads of their subjects in clamps to keep them still.

Egg whites played an important role in the preparation of an early photographic emulsion.

Cameras were once built into bowler hats, watches, walking sticks, books and even fake guns.

Colour pictures were being taken as far back as 1861.

There was a form of instant picture camera in 1864.

There have been some cameras whose shutter speeds changed with the way you held them.

The history of the camera is littered with ideas and designs that took photography forward in leaps and bounds - and just as many that took cameras one step forward, two steps back. Many of which I have tried to chronicle in the pages of this book.

The problem, of course, is that anyone who writes a book of this type immediately sets himself up to be knocked down. For every camera you mention, there's bound to be one you've forgotten or overlooked. So I make no claims about this being the *definitive* history of the camera. To produce such a book would take many more pages (and a higher price!) than we are dealing with here. Neither is this a history of photography as such. There have been many different photographic processes down the ages, many of which I freely admit you might not find in these pages.

What I have attempted here is a concise and perhaps somewhat objective history of the camera. It deals with what I consider to have been the important landmarks from the start until now. It also deals with some of those photographic processes, but only as they apply to, or are relevant to, certain stages in the development of the camera. To put the past into today's terminology: it's the hardware I'm dealing with more than the software. If it lures you into an interest in certain ages or types of camera, then there are many other excellent books around which look in greater detail at smaller areas. What I hope I have produced here is a good, all-round and well-illustrated guide to the general history of the camera.

Those illustrations have come from a number of different sources and people, to which and to whom I must pay my thanks. For many, I am indebted to Jessop Specialist Publishing who publish *The Jessop International Blue Book.* My thanks to them for allowing me to so freely plunder their files of pictures. My thanks as well to Christie's of South Kensington for further access to their own extensive files of pictures. Further credit - and more thanks - must also go to the Science Museum, *Amateur Photographer* magazine and the many press and public relations people in the British photo industry who helped with some of the more difficult to find pictures.

My gratitude must also be extended to three other people who helped with the compilation of this book, bringing me information, telling me where I got it right and - more importantly - pointing out where I sometimes got things wrong. In that respect I must first thank David Lawrence of Jessop Classic Photographica and Michael Pritchard of Christie's, each of whom meticulously read my original manuscript.

But there is also one other person, without whose original help, this book could never have been started, let alone finished. Many years ago, I belonged to a photographic society in Essex and, on the syllabus one night, there was an evening entitled 'The Story of the Camera by Dr Robert White' It was an evening which, at that time to me, promised instant boredom. Nevertheless, I went along and, entering the room, found it literally smothered in vintage cameras, all from Dr White's personal collection. Almost against my better nature, I found myself fascinated and, by the end of that evening, I was a camera collector in the making.

Since the night of that lecture, Bob White has become not only a valued friend, but also a guide and unique source of information on all things photographically antique. My thanks, then, to Bob, for access to his private collection of cameras, his library and - most of all - the information stored inside his head. Without all three, I would never have been brave enough to attempt the book you are, I hope, about to read and enjoy.

John Wade,
April 1990.

1. Things that go black in the light

TO PUT A DATE on the first camera is all but impossible. If, in fact, you assume that cameras started with the taking of photographs, we could begin with any number of different dates. It is, for instance, often suggested that photography began in 1839, the year that a Frenchman by the name of Daguerre gave the world the first viable method of capturing and fixing an image. With his announcement, came the original Daguerreotype camera, and within hours of the daguerreotype processing being announced, the manufacturers were swamped with orders. So this could be considered as the earliest commercially successful camera. It was, however, by no means the first.

There was the camera used, in 1826, by another Frenchman named Niépce who took what has come to be accepted as the first photograph. And there were others, used by scientists, inventors and amateur experimenters, even before that date. Thomas Wedgwood, for example, knew enough about the theory of photography to attempt - and fail - to take a photograph, back in 1802. Was the instrument he used the first camera? Not really, because there were cameras of a sort around much, much earlier than even that, and if we really are to begin at the beginning, we must start with the camera obscura, a device that was popular around the 11th Century, with examples actually recorded in China even as early as the 5th Century BC!

The camera obscura

The original camera obscuras were vastly different from the cameras they were later to become. For a start, you couldn't have held one in your hand; you would have had to stand inside it. They started as darkened rooms with one small hole in a wall projecting an image of the scene outside onto the opposite wall in much the same way as a pin-hole camera projects an object onto film. The only use for the camera obscuras in these early days seems to have been for observing the sun.

In 1550, Girolamo Cardano, a physician and professor of mathematics, announced that if a biconvex lens was used rather than a bare, empty hole in the wall, a far brighter and sharper image could be achieved. The idea was enthusiastically

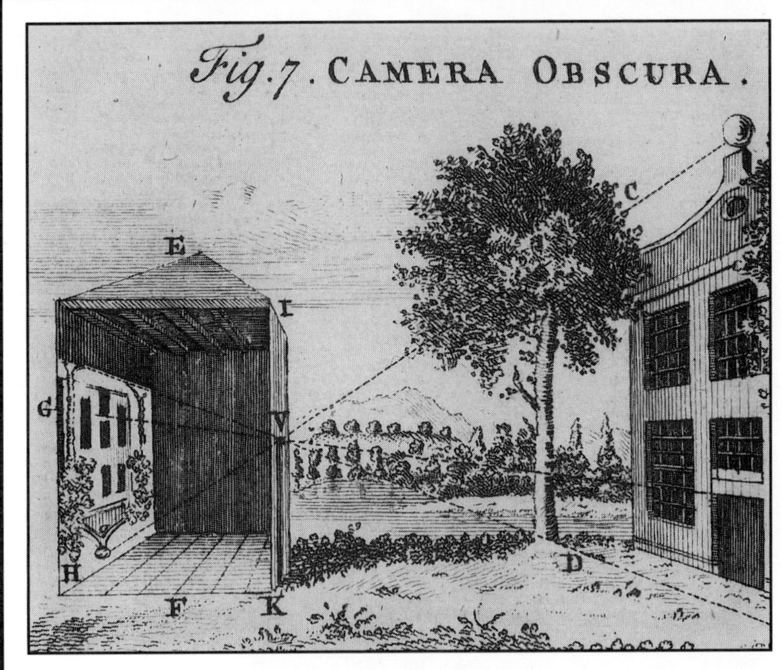

Long before photographic cameras were invented, the camera obscura was being used for scientific experiments and even entertainment. Early examples were actually portable rooms.

The camera obscura was soon taken up by artists as an aid to sketching. It became smaller, taking on different forms, like this tent type. It was still, however, a long way from being a real camera, capable of recording images.

taken up by writer and scientist Giovanni Battista della Porta, who began experimenting with various lenses until he obtained the best and sharpest image possible. Then he invited an audience to watch as a group of hired actors outside the room were projected onto the wall inside.

The reaction of what could possibly be described as the first cinema audience was not too good. The sight of the actors, ghost-like and upside-down, moving as if by magic on the wall proved too much for them. They panicked and della Porta was subsequently charged with sorcery. He was later acquitted.

From magic to science

But by the 17th Century the camera obscura had been promoted from the realms of magic to become accepted as a genuine scientific instrument much favoured by artists as an aid to sketching. Models soon became portable - in so far as a decent sized room could be said to be portable. One such model took the form of two collapsible rooms, one inside the other. The user stood inside the smaller room and viewed the scene around him back-projected through the outer wall, onto the inner wall.

Towards the end of the Century and into the early 1700s, the portable camera obscura took on many weird and wonderful forms. Some were cone-shaped, others

Before camera obscuras could be considered for photography, they had to become more portable. This example, used by Fox Talbot, shows how the artist's tool gradually began to look like a camera.

were square, but there was always a lens at one end to project the image onto a screen at the other, together with some contrivance that necessitated the artist enclosing his entire head and drawing arm in the dark behind the screen - the only way he could actually see the image in daylight.

There were camera obscuras designed in the form of tables for interior work and others built into tents for outside. There were even models designed as sedan chairs. They were, however, camera obscuras straight and simple, used by artists and destined to have little or no bearing on future photographic apparatus.

To see the point where the camera obscura first began to resemble a photographic camera, we must go back to 1676, to a point when the instrument's history came to a fork. One road led to sedan chairs and the like, the other contained Johann Christoph Sturm.

The first reflex

Sturm was a professor of mathematics at Altdorf University who described and illustrated the first reflex camera obscura. For the first time, a model was designed with a mirror at 45 degrees to the lens which reflected the image right way up onto a piece of oiled paper stretched across an opening in the camera top. This viewing screen was shaded by a box-like arrangement made of paper, into which the artist could insert both his hand and his head. It was a decided improvement on the room within a room arrangement, but its portability was still somewhat suspect.

Then, nine years later, the German monk, Johann Zahn designed a model small enough to be carried anywhere, that to all intents and purposes was a single lens reflex. It measured about 2 ft. long by 9 in. high, the lens was mounted in an adjustable tube that focused the image via a 45 degree mirror onto opal glass, flaps on the lid shielded the viewing screen and the inside was painted black to kill unwanted reflections.

For the first time in the history of the camera, there was an instrument that could be carried easily from place to place and which was theoretically capable of taking a photograph. The year was 1685 - nearly 150 years before such a thing would prove possible.

The camera owes it existence to an equal marriage of physics and chemistry, the former providing the optical and mechanical side, the latter looking after the light-sensitive materials that make photography possible; and while the physics was now ready, the chemistry still had a long way to go. Not that scientists of the time were entirely ignorant about light-sensitive chemicals. Many were aware that silver nitrate turned black when exposed to the sun; unfortunately they attributed this not to the sun's light, but to its heat.

The first man to prove that certain compounds of silver were light-sensitive rather than heat-sensitive was Johann Heinrich Schulze. He was a professor of anatomy and, purely by accident while trying to make phosphorus, he discovered that a flask filled with nitric acid and a little silver, mixed with some chalk, turned from white to a deep purple when exposed to sunlight.

Early experiments produced photogenic drawings like this.

He made stencils of words and even complete sentences which he taped to the side of his flask before exposing it to light. The result was dark words against a white background. With no way of fixing the image, however, it wasn't long before his silhouettes disappeared as the entire mixture darkened in the light.

That was in 1727 and over the next few years, a number of men became involved with photochemistry. Silver nitrate was firmly established as a light-sensitive chemical, silver chloride was also discovered to have similar properties and there followed experiments with the chemicals in both liquid form and as a coating on paper.

Towards the end of the 18th Century Thomas Wedgwood, son of the famous potter, began experimenting with sensitized paper. He made white against black copies of insects' wings and leaves by simply laying them on the paper and exposing it to light. He also managed, by the same method, to copy pictures that had been painted on glass. There was, however, no way to fix the images he obtained and they could be viewed only by candle light.

The first photograph - nearly

Undeterred by this quite serious drawback, Wedgwood next took his experiments one last and remarkably relevant step forward. He put a sheet of sensitized paper into a camera obscura, convinced that he would be able to capture its image. It was 1802 and photography was a hairsbreadth from discovery. But it was not to be. Wedgwood failed, probably through too little exposure of the material. Had he left his paper in the camera obscura longer, he would undoubtedly have obtained an image and might now be credited as the inventor of photography.

As it was, he became convinced after too short a time that the process was impractical and gave up that particular line of research. Fourteen years later, the idea was turned into a practical proposition by Joseph Nicéphore Niépce

Niépce used a camera, specially made for him. It was about 6 in. square with an adjustable tube supporting the lens. He also had a tiny camera made from a jewel box, fitted with a microscope lens, giving an image under 2 in. square and a slightly larger one fitted with a similar lens. It was with one of these that Niépce actually succeeded in producing the world's first negative which he partially fixed with nitric acid.

Fixing the image

It was a few years later, in 1819, that the celebrated English astronomer Sir John Herschel discovered the way sodium thiosulphate (which he mistakenly called hyposulphite of soda) dissolved various silver compounds and, had Niépce known of this discovery, he might have used the knowledge to permanently fix his first negatives. As it was, he abandoned the idea in favour of a different and far less practical method of recording images.

It is interesting to note here that the man who is generally credited as producing the world's first negative is Talbot. The credit is rightly attributed since it was he who first saw the practical implications of the negative as it is used in today's negative-positive system. Yet, Fox Talbot's famous 'first negative' of the lattice window at Lacock Abbey in England was made some nineteen years after Niépce produced his negatives in France.

Joseph Nicéphore Niépce

Niépce was the son of a French King's Counsellor. From childhood, he was a born inventor, on the look-out for something new with which to experiment.

Together with his brother Claude, he invented a hot air engine and even produced an early type of bicycle.

In 1816, he began experimenting with camera obscuras, trying to record the scene from his attic window using paper sensitized with silver chloride.

The reason Niépce shied away from the negative image was quite simply because it was a negative. He was after a method of recording accurate pictures of scenes from nature, and black skies with white shadows didn't answer his needs. He wanted positives, not negatives. To give credit where it is due, it must be said that he did try making positives by contact printing his negatives with fresh sheets of sensitized paper, but to no avail. The results didn't satisfy him and so he began looking in other directions to find what must then have seemed the obvious aim at the time - a method of producing a positive first time in the camera.

Real success came from his interest in lithography. Earlier experiments in this field had led Niépce to a knowledge of bitumen of Judea, a kind of asphalt whose most important properties in this context were these:-

1. It hardened when exposed to light.
2. The hardened bitumen was pale white in colour.
3. The unexposed material remained soft and could be dissolved in lavender oil.

Dissolving the bitumen of Judea in lavender oil, Niépce coated a glass plate with the mixture. A line drawing, made translucent with oil was placed face down on the glass and then exposed to the sun.

Where the light shone through the blank areas of the drawing, the bitumen of Judea hardened. Where the dark lines of the drawing shielded the sun's rays, it remained soft and therefore soluble. It could then be washed away with more lavender oil and a positive image was thus left behind. What was more, it was permanent. It couldn't fade because the sensitive soft bitumen had been washed away.

The final step

From there, it was only a short and very logical step for Niépce to try the process out in a camera obscura. Evidence points to the fact that he used his first professionally made model for the purpose, built for him by Charles and Vincent Chevalier, the famous Paris opticians, and fitted with a prism to correct the picture which would have been laterally reversed in the normal way.

For the previous few years, Niépce had been experimenting with various bases for his process and by now, having tried glass, copper and zinc, he had eventually settled on pewter. So a pewter plate was coated in bitumen of Judea, fitted into the back of his camera and placed on a window-sill with the lens aimed at the courtyard outside. It was left there all day.

At the end of the day, Niépce removed the plate. The bitumen had hardened and turned white where the light had fallen. He dissolved the remaining soft bitumen with lavender oil, leaving the pale hardened substance to portray the highlights while the

The picture that started it all. Taken by Niépce in 1826 it shows a French courtyard and took a day to expose. The picture is generally accepted as the first photograph.

base metal, darkened by treatment with iodine vapour, showed the shadows.

The process was totally impractical and the exposure time nothing short of ludicrous; but whatever failings the system had, one thing was for sure. Joseph Nicéphore Niépce had just taken what has come to be the earliest surviving fully-fixed, positive picture. He called the process heliographie. The year was 1826. And the history of the camera as we know it today had begun.

2. From today painting is dead

BETWEEN 1826 AND his death seven years later, Niépce used several cameras still in existence which were way ahead of their time. We have already mentioned the one made by the Chevaliers on which the first photograph was almost certainly taken and it was this firm who made what is now reckoned to be the earliest known camera still surviving.

This was designed on the sliding box principle; two boxes, one which slid within the other to give the degree of movement necessary for focusing. The camera measured 12 x 12½ x 15 in.with a format of 6¼ x 7½ in. The lens, no longer in existence, is thought to have been biconvex with a focal length of 300mm.

The sliding box design was one that would dominate cameras for many years to come, but Niépce also owned a crude form of bellows camera. The biggest single problem of his process was the overlong exposure times, and yet Niépce used another camera that utilized a variable iris diaphragm - not to lengthen the exposure time even more, but because he discovered that a smaller aperture would sharpen the image. Many of his cameras also had small holes in one side with a plug that could be removed for the user to observe the image as it slowly appeared on the plate.

There is no doubt at all that Niépce was a great innovator, but as far as progress was concerned, heliographie was a dead end. There was no way of improving it, nowhere for it to go; and so it fell to someone else to come up with a different process which, in its way, proved to be just as much of a dead end, but unlike heliographie, was at least practical. The process, which caught the imagination of an entire generation, was the daguerreotype process; its inventor was Louis Jacques Mandé Daguerre.

A unique partnership

Daguerre came to hear of Niépce's work through the Chevaliers who made camera obscuras for both men. When he heard that Niépce had actually managed to permanently record the instrument's image, he became intrigued because he had, for some time, been trying to do the same thing. He immediately wrote to Niépce and suggested a meeting. For a while, Niépce remained suspicious of this total stranger

Louis Jacques Mandé Daguerre

DAGUERRE was a keen and talented artist, a fact that had led to his apprenticeship as an architect. From there, he went on to become a scenery painter with the Paris Opera, progressing after a while to designing theatrical scenery.

The theatre would seem to have worked its way into his blood because he was also a ballet dancer and acrobat with a talent for tightrope walking.

Scientifically, he had no qualifications at all. What he did have was tremendous drive and unbounded enthusiasm for whatever subject caught his imagination. Diorama was a typical example.

By painting pictures and parts of pictures on a number of translucent screens arranged one behind the other, Diorama created three-dimensional illusions by the use of clever lighting. Audiences could sit in the specially built halls and watch as church interiors or exterior landscape views appeared and vanished before their startled eyes, complete with music and sound effects.

It was like magic and one of its inventors was Daguerre. With such a background, it isn't surprising that the man eventually turned his interest and remarkably talented mind to the subject of cameras.

who had approached him out of the blue, but after making some inquiries about the man, he eventually agreed to meet Daguerre. He found a man whose energy was surpassed only by his versatility.

Despite their widely different backgrounds, Niépce and Daguerre became friends and partners. The partnership was formed initially to perfect and improve heliographie and over the next four years, they each worked separately, reporting their experiments to one another by mail. But no improvements were forthcoming and all Daguerre actually added to the partnership was better camera obscuras.

It wasn't until after Niépce's death that Daguerre finally came up with his own process which, while owing much to his ex-partner's knowledge, was vastly different from heliographie - a fact that Daguerre felt justified calling it by a different name. And so the daguerreotype was born.

The Daguerreotype process

Daguerrian sensitizing box, part of the equipment needed to make and develop a daguerreotype - a process that could be dangerous to health without the right precautions.

To make a daguerreotype, a plate of copper, coated with silver, first was cleaned and polished, using pumice and olive oil.

Then, by the light of a candle, the plate was exposed to iodine crystals. This would have been in an airtight box as fumes from iodine were particularly dangerous. The fumes combined with the silver on the plate to form silver iodide, a light-sensitive compound.

When the plate had taken on a golden-yellow colour, it was ready for exposure, during which, the silver iodide was reduced to silver by the action of light.

Developing was as dangerous as sensitizing. In the latter, the photographer ran the risk of exposure to iodine fumes; in developing, he stood the chance of finding himself exposed to mercury fumes. The mercury was heated in a dish over a lamp, inside an air-tight box.

The exposed plate was now slotted face down into the box so that the fumes could rise from the heated mercury and react with the silver to form a white amalgam. Development was by inspection, through a window in the box, watching the image on the plate reflected in the pool of mercury beneath.

The plate was then removed from the box. The white amalgam made up the highlights of the picture and the salt solution was used to wash away the unaffected silver iodide so that the shadows were represented by the unchanged, polished silver beneath. Later, when Daguerre heard of Herschel's hypo, he used this in place of salt to fix his images.

The biggest factor that contributed to its success was Daguerre's discovery of the latent image; discovered, reputedly, by accident when he left an underexposed plate in a cupboard with a broken thermometer and returned to find an image had appeared as a result of the mercury fumes.

The first daguerreotypes were made in 1835, but it wasn't until later that Daguerre found a way of fixing them, using a solution of common salt.

In March 1839, Herschel, whose own experiments continually touched on light sensitive materials, presented a paper to the Royal Society with the title, *The Art of Photography, or The Application of the Chemical Rays of Light to the Purpose of Pictorial Representation.* The paper was published a week later, the first time the word photography had been used.

Daguerre's announcement

It was used again when Daguerre's process was announced to the French Academy of Sciences in July of the same year, and such was its success that artists of the time immediately began fearing for their jobs. 'From today, painting is dead,' announced the French painter Paul Delaroche when he was shown his first daguerreotype, and he meant it. For the first time, someone with no artistic ability whatsoever could reproduce scenes from nature with more accuracy than the most proficient of artists.

The daguerreotype, probably the first practical method of photography, was characterised by a positive image on a silver-plated copper base.

The Daguerreotype Camera

The original Daguerreotype Camera of 1839 was designed by Daguerre himself and built by Alphonse Giroux of Paris. It measured $19^{1}/2$ x $14^{1}/2$ x 12 in. and took $8^{1}/2$ x $6^{1}/2$ in. plates, the size which became known as whole plate only when smaller cameras were later built.

The lens had a focal length of 8 in., an aperture of approximately f/15 and focusing was by the sliding box method. Exposure was controlled by a metal cap across the lens that could be slid aside and replaced after the appropriate amount of time; hinged to the back of the camera was a mirror at 45 degrees to correct the image on the ground-glass focusing screen - although it was, of course, still laterally reversed unless a prism was used in front of the lens.

On each camera there was a metal plaque with a signature by Daguerre, guaranteeing authenticity.

Once the daguerreotype process caught on, many different cameras were made to support it. This is a half-plate version with a Petzval lens.

The daguerreotype process had arrived with a vengeance, but what of the cameras that used it? First and foremost, there was the original Daguerreotype Camera, made by Alphonse Giroux of Paris.

The Giroux camera was perfectly capable of landscape work and was in fact used for such, but the first true landscape camera was conceived by an amateur in the same year of 1839. He was Baron Pierre Armand Seguier and he designed the first daguerreotype bellows camera. In his design, two sets of bellows opened each way from a central fixed point and when folded, the camera could be stored in a large but portable box that contained all the equipment needed for working the complicated process. Seguier was also the first to put a camera on a tripod and even this could be converted into a mobile darkroom by throwing a canvas cover over the top, making a dark tent for the photographer to work in.

Disadvantages of daguerreotypes

The daguerreotype had three main disadvantages. The first was in the viewing; depending on the angle at which light struck the surface of the plate, the viewer might have seen a positive, a negative or even a combination of both.

The second drawback lay in the fact that the daguerreotype was a one-off process;

the finished picture was on the actual plate that came out of the camera and therefore could not be duplicated.

The third problem was still the lengthy exposure times that made portraiture impractical. Though this was not a fault of the process, only of the cameras of the time. The small aperture lens of Daguerre's original camera was just not gathering enough light.

It was a problem that was overcome by the insight of an Austrian professor named Andreas Von Ettinghausen, who was present when Daguerre's process was first announced. Even in 1839 he recognised the need for a better lens and, at home in Austria, he rapidly got together with Joseph Petzval, a mathematician and Peter von Voigtlander, head of the famous optical firm.

Essential accessories for the daguerreotype portrait photographer: a posing stand to clamp the subject's head and a Daguerrian posing chair.

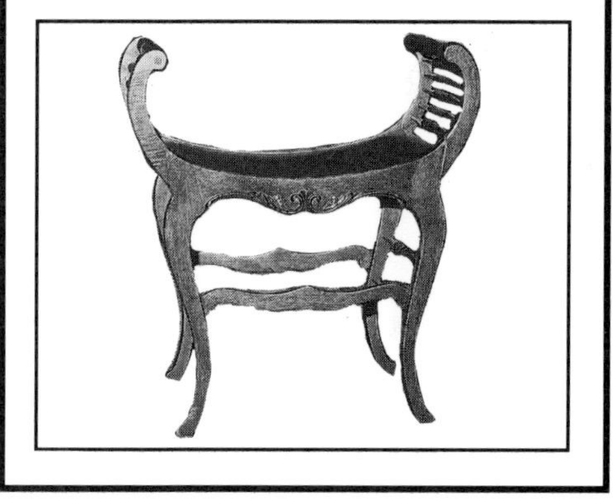

Shorter exposures

By 1840, Petzval had designed a lens with an aperture of f/3.6 which immediately cut exposure times and made portraiture a practical proposition. It was fitted to a camera built by Voigtlander, whose design owed more to the telescope than to any camera.

It was made of brass and consisted of two long cones with a focusing screen in the centre. The front cone supported the lens, the rear cone held a magnifier for inspecting the image on the ground-glass screen and the whole assembly was held in a metal cradle.

The camera was ideally suited to portraiture and yet, when Richard Beard opened the first photographic studio at London's Royal Polytechnic Institution in 1841, he used a daguerreotype camera of even more unusual design and one that did absolutely nothing to influence future apparatus. The camera, designed by Alexander

The Voigtlander Daguerreotype Camera with Petzval lens.

Wolcott of New York, used a concave mirror in place of a lens. The mirror was positioned at the back of the camera and gathered light through an aperture in the front. It reflected its image onto a plate inserted through an opening in the camera's side at the point of the mirror's focus and the sensitized side of the plate was naturally faced away from the object or person to be photographed.

Beard was the first of several leading portrait photographers to be approached by Sir Charles Wheatstone with what must have seemed an unusual request. He wanted two pictures of the same subject taken from slightly different positions so that the displacement between the two positions of the lens was approximately that of the human eyes. What Wheatstone wanted in fact was a stereo pair.

John Roberts Daguerreotype camera.

The difference between two views of a single object seen by each of the eyes was observed as far back as 280 B.C., but it wasn't until 1832 that Wheatstone found that the appearance of three dimensions could be faked by making a couple of two-dimensional drawings from different viewpoints and then arranging them in a special viewer so that each eye saw only one of the drawings.

The start of stereo

So when photography came along, the theory behind stereo pictures was ready and waiting, although, due to the design of the early viewers which let light in from all directions and the way daguerreotypes relied on the angle of light for their viewing, they didn't prove very successful in the first instance. It wasn't until the more conventional enclosed viewers were later made that daguerreotype stereo pairs became more practical.

Far better for stereo photography was the paper-based calotype picture that arrived soon after the daguerreotype. Of which, more later.

During the 1840s, as the daguerreotype process became more and more popular, different designs of camera began springing up literally all over the world.

In 1840 Charles Chevalier designed and built a collapsible daguerreotype camera with an f/5.6 lens which folded from 11 in. for its working position to 3¾ in. for carrying; in 1841, Alexis Gaudin of Paris departed from Daguerre's original size plates to produce a model taking pictures 3¼ x 2¾ in. for a camera with three aperture stops from f/4 to f/24; and in America, John Plumbe made a sliding box camera in 1842 very much like Daguerre's original model but smaller, only 5¾ x 6 x 4½ in.

Like much that had gone before and much more that was still to come, Daguerre's

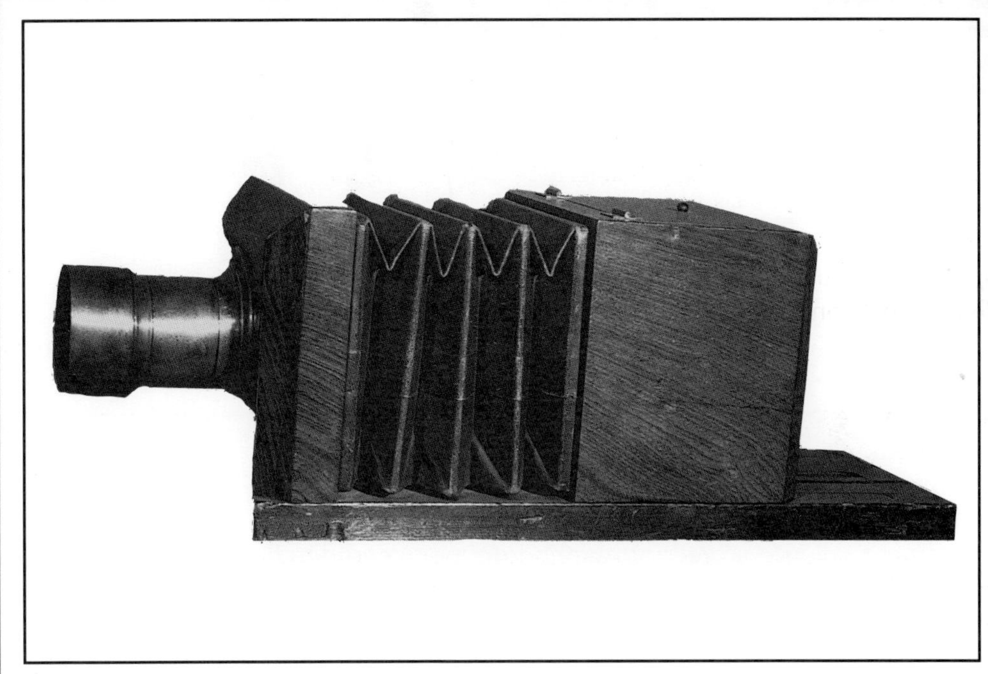

Most early daguerreotype cameras used the sliding box principle for focusing, soon to be replaced by the more compact bellows method. This Lewis Style Daguerreotype Camera was one of the first American models with bellows. It was made for quarter-plate images.

invention was clumsy and, in some respects, a positive danger to health. Its saving grace was quite simply that it worked and its biggest attraction was its use for portraiture. In an age when the only previous method had been painting or drawing, a person could now have his likeness reproduced by the far quicker and relatively inexpensive daguerreotype.

Customers flocked to studios from far and wide as the family portrait suddenly ceased to be the prerogative of the rich. Portraiture for the masses had arrived at last.

3. The day of the mousetraps

NIEPCE TOOK THE first photograph, Daguerre adapted the process to his own ends and made it far more practical, yet the man who is probably credited more than any other with being the father of modern photography is William Henry Fox Talbot.

When the daguerreotype process was made public in 1839, the news must have brought deep disappointment to Talbot. For the previous five years, he had been working to perfect a photographic process of his own and he had every hope of being the first to announce a practical method of photography to the world.

Talbot had been led to his photographic experiments, like so many before him, by a fascination for the camera obscura and another artist's aid of the day known as a camera lucida. His early experiments were along similar lines to those of Thomas Wedgwood, producing silhouettes of leaves, lace and feathers on sensitized paper - no developing, just exposing until the image physically appeared. Unlike Wedgwood, however, he found a way of fixing his images with a solution of salt.

This breakthrough came because Talbot coated his paper with sodium chloride (common salt) before applying the usual silver nitrate. This gave him a coating of silver chloride which he hoped would be more sensitive than the nitrate compound. In fact, it showed him that the more salt he used, the slower was the speed with which the paper darkened in the light, a fact that led him straight to the use of salt water as a fixing agent.

Negatives and positives

But negative images, even if they were fixed, were not enough to satisfy Talbot. He soon took his process a step further and in so doing, he made history. Because rather than search for a way of making positives in the camera the way others had, Talbot paved the way for the negative-positive system of today's photography by contact printing his negative picture with a fresh piece of sensitized paper, using the sun as his light source.

The natural and, by now, obvious last stage was to try fixing the image from a camera obscura. His initial experiment failed through lack of adequate exposure,

William Henry Fox Talbot

Talbot was a man of extraordinary vision. He was Squire of the Manor in the Wiltshire village of Lacock and his talents took in much more than the early photographic experiments with which he is more usually associated.

A brilliant mathematician, researcher into light and optics, the inventor of the polarizing microscope and a finger in every pie from English Literature to astronomy, taking in botany on the way, that was Talbot. He wrote three volumes on etymology, published the first book to be illustrated by photographs and was a Member of Parliament for a few years.

Even at the age of seven, his diary showed he was experimenting with optics and later, at Harrow, he actually caused an explosion which had him banned from active experimentation. After Harrow, he went to Trinity College, Cambridge, but it was in his thirties that his creative work began.

During the decade that spanned 1830 to 1840, he flirted for a while with politics, married his wife Constance, became involved in chemical, physical and mathematical research, published the three etymological volumes and finally began his researches into photography.

Three cameras, used by Talbot which, because of their size, earned the name 'mousetrap cameras' from the photographer's wife.

something which he was swift to recognise; but while others had concentrated on building bigger and better lenses, Talbot turned in a different direction. He started making smaller cameras.

Small cameras with short focal length lenses meant that the light gathered by the lens could be concentrated on a small area, rather than spread over a large one, resulting in a brighter image and consequently shorter exposure times. The cameras, made by a village carpenter and fitted with microscope lenses, were no more than 6cm square, giving images 1 in. square. They sat quite comfortably in the palm of the hand, a fact that seems to have endeared them to Talbot's wife, Constance: for it was she who gave them the name that passed into history. Because they were so small, she called them mousetrap cameras.

The earliest negative

It was with one of these in 1835 that Talbot made the earliest negative still surviving. The picture was one of a lattice window at Lacock Abbey, taken from inside looking out and exposure time was within the region of half an hour.

A positive made from the earliest surviving negative, made by Talbot with his photogenic drawing process.

Talbot's calotype process, used to document the building of Nelson's Column in London's Trafalgar Square.

The process itself, which still relied on the negative material being exposed until an image actually appeared upon it, was known as photogenic drawing and it never really caught on seriously. The latent image, whose discovery would have cut exposure times so drastically was still unknown to Talbot and during the next four years, he concentrated on adapting photogenic drawing to various subjects that ranged from the exteriors of buildings to pictures through a solar microscope.

Larger cameras

It wasn't until 1839, when news of Daguerre's process reached his ears that Talbot began trying to speed up his own process. The mousetrap cameras, he realized by now, were defeating their own object. Their image was brighter, the exposure time was less, but with no way of enlarging in those days, who wanted a picture only 1 in. square?

During 1839, Andrew Ross made some larger cameras for Talbot that were fitted with achromatic lenses and it was while using these that he had a lucky accident that changed the course of his experiments overnight.

The calotype process

Using a candle for illumination, a sheet of good quality writing paper was washed, dried and then sensitized by first floating it in silver nitrate solution. This took about three minutes and the resultant coating was allowed to dry before soaking the paper again in potassium iodide solution. The two chemicals reacted to produce the light-sensitive silver iodide. In this stage, the paper could be kept until required.

When a photograph was to be taken, the paper was brushed with a solution of silver nitrate, acetic acid and gallic acid to make it more sensitive, partially dried and used in the camera still damp. The exposure was then made - anything between one and two minutes.

After exposure, the paper was again treated with silver nitrate, acetic acid and gallic acid, then warmed in front of a fire for a few minutes when the image finally appeared. It was fixed, in the early days with potassium bromide, but later, when Talbot learnt of Herschel's discovery, he too used hypo for the purpose.

For printing, silver chloride paper was used, prepared by floating the paper first in ammonium chloride and then in silver nitrate. The paper negative was placed in contact with this, emulsion to emulsion and exposed to the light of the sun until the positive image appeared on the new sheet of paper. It was then fixed in the normal way.

A finished calotype, because of the paper's texture, ended up as a soft-looking print with slightly fuzzy edges. It never did acquire the sharp definition of the daguerreotype but in many ways, especially in portraiture and landscapes, its very imperfections were a major part of its charm.

Having underexposed some sheets of sensitized paper during the course of one experiment, he put them in a cupboard with the thought of using them again. When the need arose, he took the paper from the cupboard and began resensitizing it. To his astonishment, an image appeared - a picture of the scene that he thought to have been under-exposed. And so, like Daguerre before him, Talbot discovered the latent image purely by accident.

It was 1840 and for the first time, with exposures down to as little as three minutes, Talbot had a method of taking portraits. What started as a lucky accident ended as a whole new photographic process whose name was taken by Talbot from the Greek word *kalos,* meaning beautiful. He called his pictures calotypes. Let's look at the cameras that used the process.

We have already mentioned the mousetrap cameras and even when Talbot had larger cameras made, the early designs followed the cube-shaped designs. Those used for photogenic drawing had plugs in the lens panels that could be removed to see how the image was progressing; but later, when Talbot discovered the latent image and exposure times became shorter, his new cameras were built without these spy-holes as they would have admitted too much light and spoilt the picture.

The early cameras had lenses held in adjustable focusing tubes and one had a ratchet that wound the camera to any degree of tilt that the user might require to take his picture. Another of his cameras was made of deal with a back that curved the paper to allow for the lens's curvature of field.

More sophistication

Most of these cameras, however, were really quite crude - no more than a box with some device for altering the distance between lens and sensitive paper. But with the end of photogenic drawing and the beginning of the calotype process, Talbot began looking towards more sophistication in his cameras. He bought a Giroux-made

Cundell's calotype camera

George Cundell was a gifted amateur who, in 1844, suggested a calotype camera designed on the sliding box principle without a baseboard, just two boxes, one which slid within the other. It boasted three firsts: a focusing scale, a lens hood and internal baffles to cut down reflections inside the camera.

The focusing scale, however, was there for more than the expected reason. The camera was fitted with a simple meniscus lens that was prone to all the aberrations of its type, not the least of which was the fact that rays from the violet end of the spectrum came to a point of focus before those from the red end. And, as it was the violet rays that had most effect on the photographic emulsions of those days, it was possible for an image that appeared sharp as a pin on the focusing screen to be just off focus on the negative.

Cundell got round this problem by fitting his camera with not one focusing scale, but two.

The image was focused on the ground-glass screen, a number was read off on the first scale and then focus shifted so that the corresponding number was now adjusted on the second scale. The result was a perfectly focused picture every time.

The early art of portrait photography, using Talbot's calotype process in an outdoor studio.

daguerreotype camera which he adapted for use with calotypes. For portraiture, he used a fixed focus, all metal camera made by Lerebours of Paris. The camera, which had a choice of three apertures on a sliding disc in front of the lens, was in the form of a metal cylinder attached to a square holder for the sensitized paper.

In 1845, the same company made a sliding box camera for calotypes 3 x 3$^{1}/2$ in. This too had a choice of apertures on a rotating disc and the ground-glass screen was shielded by a folding wooden hood that took the place of the more traditional black cloth.

Talbot also used a 2$^{1}/2$ x 3 in. camera fitted with a special lens made by Chevalier and which was supplied with two different front elements - one for landscapes, the other for portraiture.

Yet despite the fact that photographic processes were changing and improving all the time, the 1840s were surprisingly devoid of any real revolution in camera design. Most of the cameras used by Talbot and his contemporaries were no more than adaptations or slight improvements on those used by Niépce thirty years before. One of the few exceptions and innovations was the camera designed by George Cundell.

But generally speaking, it wasn't until the 1850s, that camera manufacturers began getting more adventurous with their designs and then they turned their thoughts towards making models with the accent on convenience. Lighter cameras, models that were easy to carry around became the order of the day.

New camera designs

An early form of magazine camera made its appearance in 1850, designed by Marcus Sparling. In this, ten sheets of paper could be stored in separate holders ready for use and, after exposure, each sheet dropped down into a compartment under the camera.

In 1851, Richard Willats came up with a design whose body was made, not of wood or metal, but of cloth. The camera back could be clamped into a sliding plate in the base and was separated from the lens panel by the conical-shaped cloth body that expanded and contracted as necessary so that lenses of various focal lengths could be used. The result was a camera that took pictures as large as $10^1/2 \times 8^1/2$ in. but collapsed to only 4 in. deep when folded.

In 1853, there was another magazine camera devised by George Stokes which could take up to twenty calotypes in rapid succession; but one of the most ingenious devices of the decade came from A.J. Melhuish and J.B. Spencer who, in 1854, dreamt up what must surely have been the first spool of roll film. It was made up of sensitized wax paper, rolled onto spools of different sizes and fitted into a special holder that could be fitted to the back of a camera.

But we're getting beyond ourselves. Back in the 1840s, if cameras weren't so inventive, processes were. Although Talbot is the man most associated with paper negatives, it must be noted before leaving the era that he was by no means the only scientist in the field. From the moment a practical method of photography was first announced, experimenters everywhere began jumping on the bandwagon.

Dedicated scientists, out and out amateurs, everyone from vets (Friedrich Gerber) to clergymen (The Rev. J.B. Reade) started claiming and showing proof of prints on paper and even some on leather. Sir John Herschel, within a week of hearing about Daguerre's success, succeeded in making a picture on paper sensitized with silver carbonate and, of course, fixed with hypo - his own discovery.

One of the most significant advances was made by a French civil servant named Hippolyte Bayard whose process, if it had been given better recognition, could have been a strong rival for the daguerreotype. In 1839, he made photogenic drawings

The Pencil of Nature was the first book to be photographically illustrated.

similar to Talbot's, but on hearing about daguerreotypes, he turned his attention to making direct positives in the camera. He succeeded too, but didn't make his methods known until 1840.

Paper positives

Briefly, his process was this: Making silver chloride paper similar to Talbot's, he then blackened it by the action of light before soaking it in potassium iodide and exposing it in a camera obscura. The process worked because when light struck the paper after its final coating, iodine separated from the potassium iodide and bleached the blackened paper, thus giving white highlights and leaving the already black paper for the shadows.

Yet despite the fact that so many were producing paper prints - some even before

Talbot had produced his - the credit still rests firmly with him. Perhaps it is because Talbot was one of the few to recognise the necessity for having an intermediary negative and a latent image. While others shunned negatives because they were negatives and not positives, Talbot saw the importance of being able to recreate the same picture over and over again from one master negative and it was this that no doubt led him to produce the first photographically illustrated book.

Pencil of Nature, written and illustrated by Talbot was published over the course of two years beginning in 1844. It comprised six volumes and the complete set contained twenty-four original calotypes, each one pasted in by hand. Two hundred and seventy-four copies of the first volume were produced, falling to 73 of the last. The prints were made at a special printing works cum studio set up in Reading where paper negative originals were used to mass produce the illustrations.

The fixing of the finished calotypes, however, did not seem to have been carried out too well and many of the prints faded soon after the books were bought. It was criticism resulting from this fault that probably led Talbot to turn his talented mind to producing a more reliable method of mechanical photographic printing and by 1851, he had developed a photographic engraving process which was basically similar to that still used today and known as photogravure.

But that's another story.

4. Have darkroom, will travel

OF ALL THE WEIRD and wonderful concoctions dreamt up by inventors to help photography along its way, one of the strangest must have been a discovery made by Niépce's cousin, Abel Niépce de Saint-Victor. His rather curious finding was a photographic emulsion that used as its base the freshly-beaten white of a new-laid egg.

The result was the albumen process with which Saint-Victor took what is often credited as the first photograph on glass. Not so. True, his was the process which popularized glass plate photography, but it was a long way from being the first. Niépce used the bitumen process to take pictures on glass in 1829, as did Daguerre a few years later. And a decade before Saint-Victor's discovery, Sir John Herschel used a coating of silver bromide (sometimes iodide of chloride) on a glass plate, then washed it over with silver nitrate immediately before exposing it in the camera still wet. The resultant negative plate could be used either for printing positives onto paper or backed with a black background to give the appearance of a positive (the exposed emulsion reflected the light to show highlights, while the dark background showed through the clear glass to represent shadows.) Herschel used this latter method in 1839 to produce a picture which today is reckoned to be the earliest surviving photograph on glass.

The albumen process

In theory, Herschel's glass plates should have given far better positives than Talbot's paper negatives and yet it was the calotype that grew in popularity and glass plates took a back seat until 1847 when Saint-Victor arrived on the scene with his albumen process. This is how it worked:-

The white of a fresh egg was beaten until stiff when potassium iodide was added. The thick substance was then spread onto a glass plate and sensitized in a solution of silver nitrate and acetic acid. After exposure, the plate was developed in gallic acid, fixed, dried and printed in the same way as a calotype.

Photographers in general, however, weren't too enthusiastic. The process was very good for recording fine detail and the finished prints were better than the average calotype; the drawback was in the fact that exposures amounted to anything up to fifteen minutes. The system was too slow and depended, believe it or not, on the freshness of the eggs.

Even so, the albumen process did remain popular for some years in another form. The slowness that ruined its chances of success on plates didn't matter nearly so much when it was adapted for use on printing paper, and albumen paper remained popular as a substitute for silver chloride paper until the end of the century.

Collodion and doctors

Meanwhile, an important discovery was being made in a different field by Dr J.P. Maynard of Boston, USA. He discovered that when guncotton was dissolved in a solution of ether and alcohol, it produced a viscous liquid that dried into a hard, transparent film. He gave the substance the name collodion, but having discovered it, he could find no practical use for it. Doctors of the time, however, were ahead of him and used its hardening properties for dressing wounds.

It fell to Frederick Scott Archer, to see the value of collodion for photographic purposes. His discovery was the wet collodion process and, with its announcement, exposure times were suddenly cut to as little as five seconds. Photographers, however, had to pay for

Frederick Scott Archer

Archer was the son of an English butcher, a coin expert turned sculptor who learned the calotype process so that he could take photographs of customers who wanted their busts modelled. Soon, the photography side of his work began to interest him more than modelling and it wasn't long before he started experimenting.

His experiments led him to the use of collodion, but whereas others had used the substance because of its properties when dry, Archer took the unusual step of using it wet - and in so doing, he created a minor revolution in the history of the camera.

Archer perfected the process in 1850 and published its details a year later. Wet plate photography was destined to remain popular for another twenty years.

One example of the kind of mobile darkroom used by wet plate photographers.

this luxury with the effort needed to actually use the process. The plates had to be used wet and therefore needed to be made within minutes of their exposure in the camera. And for that, the photographer had to take a portable darkroom everywhere.

In 1853, Archer himself designed a camera which was also a darkroom. By means of elasticated sleeves on its side, the photographer could carry out the complicated sensitizing and developing process actually inside the camera as well as making the exposure. But more usually, darkrooms were separate from the camera - sometimes carried on the photographer's back, occasionally mounted on a horse-drawn cart, but more often than not, loaded onto a sort of gigantic wheelbarrow.

And if all that trouble for just one picture seems impossibly complicated, remember that the Crimean War and the American Civil War were both photographed by this process. Also, David Livingstone went on his famous travels with a complete wet plate outfit.

The new method of photography on glass, despite its clumsiness, soon caught on.

Collodion wet plate process

The photographer who used the wet plate process first had to make his collodion. This he did by dropping cottonwool into a mixture of nitric acid and sulphuric acid which, after drying, gave him guncotton. He dissolved this in ether and alcohol to make collodion. Armed with this and much more besides, he loaded up his mobile darkroom and ventured forth.

Finding his subject, our photographer retired to his darkroom where he took a piece of glass and washed it thoroughly in alcohol and powered pumice. A mixture of collodion and potassium iodide was then poured over the glass, tilting it carefully until the whole surface was evenly covered. Then, by the light of a candle, the plate was lowered into a bath of silver nitrate and water. Five minutes later, the plate was removed and examined. Only when it was a creamy white all over was it ready for the camera.

It was then exposed, still wet, and then taken back to the darkroom for development with pyrogallic acid before the collodion dried. It was fixed and washed in the normal way and the final print made on silver chloride or albumen paper.

Daguerreotypes and calotypes that had previously been competing with one another suddenly found themselves together in competition against the new wet plate process and it soon became obvious which of the three was going to win the day.

The cameras were ready and waiting. Any model that had been used for daguerreotypes or calotypes was perfectly suitable provided it was used with a wet plate dark slide, and all that needed was four silver wires across the corners to keep the plate in place and a gulley at the base to catch the inevitable drips.

Wet plate camera designs

The cameras that grew up around the wet plate era are difficult to put into any convenient category. They ranged from a miniature built by T. Morris in 1859 which measured a mere 1½ x 1½ x 2 in. and took portraits for lockets less than 1 in. square, to a camera built in 1860 by a Scottish amateur named John Kibble. This one was enormous. The plates measured 110 x 90 cm and the entire apparatus was mounted on a horse-drawn cart. Perhaps the size of the plates had something to do with Kibble's profession: he was a builder of greenhouses.

But despite the size of Kibble's camera, more and more people were turning their minds to making models more compact and portable. With so much to carry around in the way of chemicals and darkroom, the last thing they wanted was a bulky camera. Now the thing about all cameras is the fact that no matter how sophisticated they may be, they still represent what is essentially an empty box. So what could be more logical than to build a camera which, when not in use, could be folded flat.

In 1853, Thomas Ottewill of London made a 10 x 8 in. plate camera for landscape work in which the lens panel and focusing screen could be removed and the sides folded inwards in a V shape, to make a flat package 20½ x 12¾ x 3 in. In its working position, the camera was a sliding box design with a lens panel that slid vertically or horizontally - an early form of rising and crossing front.

Probably the biggest single revolution of the 1850s as far as camera design was concerned, was the introduction of the bellows. Not that there was anything really

One camera, four lenses to produce four images. This is the American-made Anthony Four Gem Carte-de-visite Camera. It was a wet plate outfit on a cherry wood stand, made around 1860. Pictured beside the camera and stand is a wet plate holder, dipping tank and a box of early colouring dyes.

new about putting a folding leather bag between the lens panel and the plate; Niépce owned a crude version of just such a camera and there was a daguerreotype bellows camera back in 1839. But the wet plate era saw the bellows camera become a serious commercial proposition.

A good example is the 1851 model made by the father and son team of William, William H. and Henry J. Lewis who were responsible for the first commercially produced bellows camera in America. It was, in fact, designed for the daguerreotype process and was consequently known as the Lewis Daguerreotype Camera, but it could of course equally be used for the wet plate process.

The Lewis camera measured 6¼ x 7 x 16 in. and had an f/4 lens of 6 in. focal length. It was made in two parts, a fixed lens panel linked by square bellows to the rear section, which took the form of a box capable of sliding backwards and forwards in a grooved base. The bellows were leather-covered fabric and the rest of the camera was made of veneered wood.

Traditional and untraditional ideas

One early British bellows cameras was designed by the architect Captain Fowke and made by Ottewill in 1856. Front and back were joined by bellows, together with a brass rod to hold everything square. But for some reason best known to himself, Fowke designed a pyramid-shaped wooden box to hold the lens that stood out from the front of the bellows like a sore thumb. The result was both unnecessary and ungainly.

It was the time when cameras began to take on the traditional mahogany, leather and brass look which so typifies an old camera in the minds of most people; but there were still one or two models that were totally unique in their design and the Pistolgraph was a perfect example. First developed by Thomas Skaife in 1856, the Pistolgraph was made of brass and took miniature plates just 1 in. square.

The Pistolgraph, made by Thomas Skaife in 1856.

The camera took the form of a brass tube made in three telescopic sections for focusing and moulded to a square plate holder. Unlike many of the cameras of that time, the Pistolgraph had a shutter worked by rubber bands which gave an exposure of about 1/10 sec.

The Dubroni camera

The Dubroni, complete with rubber bulb that introduced chemicals into the camera body for both sensitizing and developing a wet plate. Was this the closest thing to an instant picture camera at the time?

One of the more unusual cameras of the time was the Dubroni whose name was an anagram of its inventor, a Frenchman named Bourdin.

It comprised a camera and darkroom all in one; by no means the first, but definitely the neatest and most practical. The Dubroni first appeared in 1864. To look at, it was a wooden box measuring 2³/4 x 3¹/2 x 3¹/3 in. with a lens on the front, a plate holder at the back and a rubber bulb on top. Inside, there was an earthenware container.

A collodion-coated plate was placed in the back and then the sensitizing solution introduced into the camera and onto the plate by means of a tube from the bulb. The camera was then tilted until the plate was evenly covered.

After exposure, developing, washing and fixing were also carried out inside the camera, using the tube and rubber bulb to insert and extract the solutions. The photographer could inspect the progress of his developing by holding the lens up to his eye and illumination for this came from a red window in the rear of the camera which, of course, had to be covered at the time of exposure.

An English-made wet plate stereo camera from around 1858.

What gave the camera its unusual appearance though was the fact that the whole thing was mounted on a ball and socket head which, in turn, was fixed to a small wooden box that could be used for storage of plates etc.

Stereo cameras steadily grew in popularity during the wet plate era with models like J.B. Dancer's Stereocamera of 1856. This was a British sliding box camera designed for wet collodion plates but with a number of original features that equally allowed it to be used with collodio-albumen dry plates (dealt with in the next chapter). It became one of the earliest magazine cameras in which plates could be carried actually within the design. These were housed in wooden frames that rested in a box attached to the camera's base. They were lifted into the film plane by a metal rod that screwed into the top of the frame.

Each lens of the camera was covered by a circular disc with aperture stops graded one to five. A metal bar, pivoted at a point between the lenses, blocked the light from each and swung aside with a see-saw movement to form a crude type of shutter.

But stereo cameras of the time didn't necessarily have two lenses. The Moore Single Lens Stereocamera of 1865 for instance was basically a wooden box camera with a repeating back for taking two shots on a single plate. The whole assembly was mounted on a track fixed to the top of the camera's case. Between the two exposures, the camera was slid from one end of the track to the other and the repeating plate back slid into its alternative position. The result was a stereo pair on the same plate.

Two cameras or two lenses

There were even stereo set-ups that consisted quite simply of two identical plate cameras mounted side by side and set at the right angle to produce a stereo pair, one picture from each camera. Most stereo cameras of the time, however, used two lenses.

Stereo cameras aside, the wet collodion process lent itself basically to a number of different end products, apart from straightforward prints. The ambrotype was one

The Ambrotype

An ambrotype was quite simply a wet plate negative that had been fixed with cyanide rather than hypo, treated with nitric acid or mercury bromide and then mounted against a dark background.

The original dark tones of the negative were bleached by the treatment to an equivalent tone of white and the black background showed through the clear parts of the negative. The resulting picture was therefore seen as a positive.

A one-step collodion positive that didn't rely on the bleaching stage could also be made by developing the plate in iron sulphate, a chemical that gave a white image rather than the usual black. Mounted against a dark background, the end result was another form of ambrotype.

But like the daguerreotype, the ambrotype was another one-off process that used the plate from the camera for the finished picture and it gradually declined in popularity as the carte-de-visite craze swept through the photographic scene.

Two cartes-de-visite, a popular product of the wet plate era.

such. Although looked down upon by professional photographers who considered it to be nothing more than a cheap and cheerful imitation of the daguerreotype, the process became very popular for a while from about 1852 onwards.

Carte-de-visites

The thing that more than anything else killed the ambrotype was the carte-de-visite, which rose to popularity during the 1860s. It offered portraits $3^1/2$ x $2^1/4$ in. mounted on a card 4 x $2^1/2$ in. and used by friends as a personal kind of visiting card; hence, carte-de-visite. These were often taken on multiple lens cameras that took as many pictures on one plate as the camera had lenses, and sometimes the addition of a repeating back doubled the amount of pictures that could be taken at a single sitting.

It was an idea that caught on initially during the daguerreotype period when it was impossible to duplicate pictures and so getting a number of portraits on one plate was a distinct advantage. But even with wet plates, the actual process of producing prints

The Sutton Panoramic

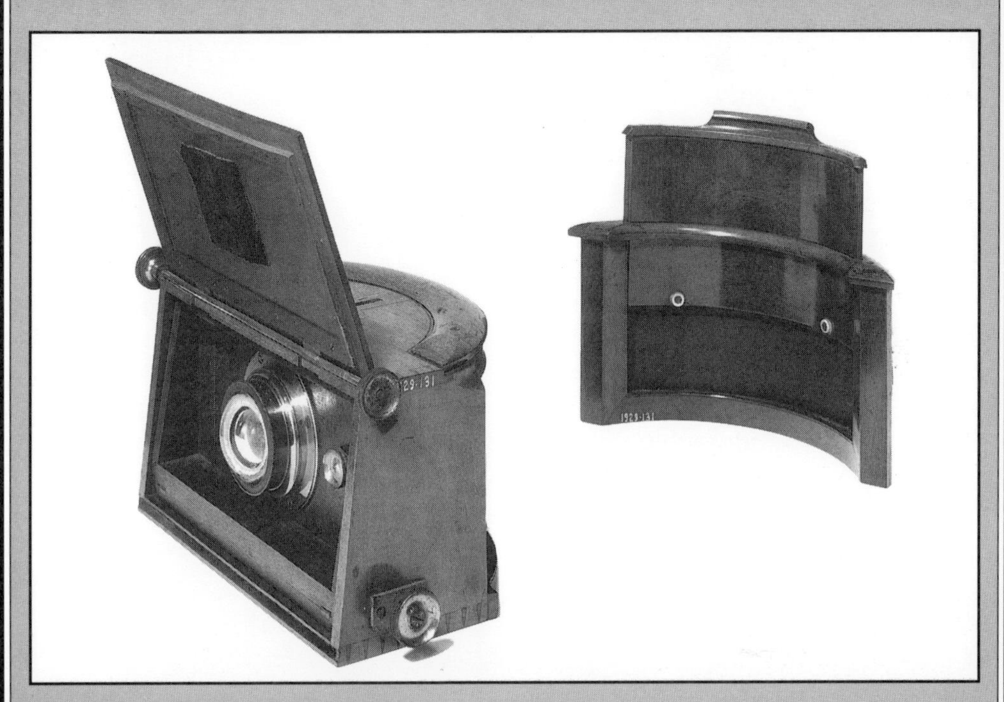

The Sutton Panoramic camera was one unusually interesting model from the wet plate days. It was made by Ross and patented in 1859. The factor that set it aside from other cameras of its time was the fact that it used a lens filled with water.

Thomas Sutton's Panoramic Water Lens was, in fact, a glass sphere filled with liquid and fitted with a central stop of f/12.

It covered a panoramic field of view of 120 degrees and, to compensate for the lens's curvature of field, the camera back had to be curved. A camera with a curved back, of course, presented no problems when flexible film was being used, but for a plate camera, things were a little more complicated.

So the Sutton Panoramic used, not only a curved focusing screen, but also curved plates and - since the camera was built for the wet plate process, even a separate curved tank for sensitizing them.

from negatives was considered time-consuming and the fact that a number of identical portraits could be produced by contact printing one plate was considered a boon.

So the 1860s saw the popularization of cameras that sported two, four, even eight lenses. There was one, called the Dragon Microphotographic Camera that took 450 minute exposures on $3^1/4$ x $1^3/4$ in. wet plates by means of twenty-five lenses in combination with horizontal and vertical shifts of the front and back of the camera. This unique piece of equipment was, however, used for re-photographing negative plates to give a series of minute positives for mounting together with a viewing lens in a piece of jewellery or individual tiny viewer. Today, these minute pieces of photographica are known as Stanhopes after the Stanhope lens to which the microscopic pictures were fixed for viewing.

Ferrotype cameras

Multiple lens cameras were equally popular with ferrotype photographers for a process which was yet another variation on the wet collodion system. The ferrotype's nearest relation was the ambrotype and it was, in fact, introduced shortly after its glass cousin. It never did really achieve the popularity that the ambrotype enjoyed at its peak but, unlike the ambrotype, it held on for many years to come and ferrotype photography was still being practised long after the ambrotype was dead and gone.

The ferrotype was a form of positive collodion plate, but instead of being made on

The Anthony Tintype camera and a ferrotype from this period.

glass and mounted on a dark background, it was made directly onto black-enamelled tin. The process was cheap and nasty, the image vastly inferior to both the ambrotype and the daguerreotype that it vaguely resembled, but its virtues were that it was both quick and cheap, and it was soon taken up by seaside and carnival photographers who very often used cameras that were combined with processing units to give an on-the-spot portrait to customers in a matter of minutes.

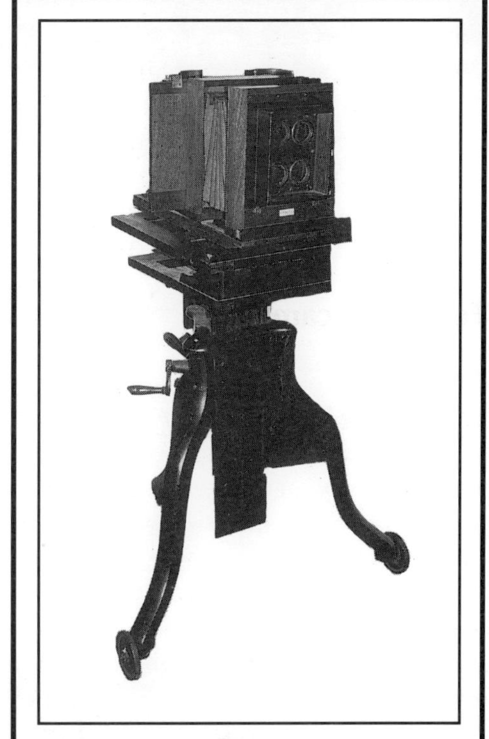

Multiple lens cameras were popular at this time. This is the Anthony Victoria Ferrotype camera.

Cameras for the ferrotype process ranged from the multiple lens models like the Lancaster Postage Stamp Camera of 1896 that took six small, square pictures, to others such as the single lens model made by Jonathan Fallowfield in 1890. This one took ferrotype pictures $2^1/2 \times 1^3/4$ in. which could be developed immediately after exposure in a tray at the base of the camera. It had an f/3.7 lens and a roller blind shutter.

The Aptus Ferrotype Camera made by the Liverpool firm of Moore and Co. in 1895 had a magazine on the base that held up to 100 sheets of sensitized tin plate. These were taken one at a time by a lever equipped with a suction device and held in the film plane for exposure, then dropped straight into a tray of developer.

The Aptus perfectly illustrates the strange appeal of the ferrotype process which never became truly popular and yet just refused to die. Because, strange as it may seem, there were models of the Aptus, not too different form the 1895 ancestors, still being built and sold by Moore and Co. as late as 1955.

5. Photographic bowler hats

WE NOW ENTER a period in the history of the camera that in many ways reads like an extract from a cookery book. We've already seen how new-laid eggs played their part in photography, now consider the following edible ingredients that found their way into the darkroom: beer, caramel, coffee, ginger wine, honey, liquorice, malt, milk, sherry, sugar, tea, treacle and vinegar.

All these and more besides were used by researchers in an effort to get away from the cumbersome wet plate system and find something that would preserve sensitized plates in a dry or semi-dry state long enough for the photographer to be able to take pictures away from his darkroom.

The first practical dry collodion process came from a French chemist named Dr. J.M. Taupenot in 1855. His was the collodio-albumen process which relied on coating a collodion plate with iodized albumen and then dipping it in silver nitrate, thus giving two sensitized layers. In this stage, the plate could be dried and kept for several weeks before use. The convenience, however, was somewhat negated by the loss of sensitivity; the dry plates were up to six times slower than their wet equivalent.

During the course of the following decade, numerous photographic processes came and went. Hardly a month passed without someone coming up with a new idea that just as quickly died.

But one thing was for certain: despite their lack of speed, dry plates were beginning to catch on and it wasn't long before a company was set up for the purpose of making and selling them to the public. For the first time, the photographer had no need to be a chemist. He could take a photograph without having to make his own plate beforehand. Instead, he could buy them dry over the counter and use them as and when needed for up to six months.

Dry Collodion Plate Company

George Eastman is the man often credited with starting the first commercial dry plate company, but twenty-two years before Eastman began his business in 1880, an Englishman, Dr. Richard Hill set up the Patent Dry Collodion Plate Company in

An early dry plate example - the New Model camera, made around 1895.

Birmingham and, by the end of 1858, he had forty agents throughout England and Canada who stocked and sold his plates.

And a camera for the process? How about the slightly unusual Photo-Binocular of 1867. Made in the style of a pair of binoculars with one lens as camera and the other as viewfinder, the camera held a drum-shaped magazine with fifty dry collodion plates inside.

This magazine could be clipped to the camera immediately prior to exposure when a plate was dropped into the plane of focus. The magazine was then removed, the exposure made and the magazine clipped on again to drop the plate back into its receptacle. A knob on the side of the drum brought the next plate into position for loading again.

The only drawback was the lengthy exposure time. Since the camera looked like a pair of binoculars, you would be forgiven for assuming it could be hand held. But you'd be wrong. The Photo-Binocular had to be used on a tripod which rather defeated its own object.

A major advance came in 1864 with the first commercial emulsion; silver bromide already mixed with collodion so that photographers could make their own dry plates by simply pouring the mixture over a sheet of glass and using it dry. For the first time, no silver bath was needed for sensitizing.

And yet, even with all the progress that was being made, photographers were still basically on the wrong track. Although the plates being made both professionally and by amateurs were now dry, they were still no more than dry equivalents of wet plates; they still relied for their success on collodion. And they were still very slow.

Then came the innovation which, in many ways, was the last, because it marked the start of a system we still use today. The man responsible was Richard Leach Maddox; his invention was the gelatine dry plate.

Gelatine, a jelly-like substance made from the bones and hides of cattle, had actually been used by Hill in his commercially produced plates, but only in association with collodion.

Other people, including Saint-Victor, had experimented with the substance, but to Maddox goes the credit for making the process work. Whereas so many before him had used gelatine as an aid to preserving collodion, Maddox took the brave step of using it instead of collodion.

His emulsion consisted of gelatine containing nitric and hydrochloric acids, cadmium bromide and the ever present silver nitrate. It was coated onto glass, used dry and developed in pyrogallic acid and silver nitrate.

With this almost deceptively simple method of plate making, Maddox changed the history of picture-taking, and yet the strange thing was that he seemed to lose interest in the idea as soon as it was made public, leaving any improvements that could be made to others. Two years later, in 1873, John Burgess, an English photographer, adapted Maddox's methods to his own ends and offered the first commercial gelatine process for sale. Photographers bought his product, made their own plates with it and, for the first time found they had a dry plate with the speed of a wet plate.

Mass production

From there, it was only a short step to turn the production of gelatine dry plates into a commercial enterprise. By 1878, four English manufacturers were offering the new plates for sale and it wasn't long before the news spread across the Atlantic. In America, a 24-year-old bank clerk began experimenting with photographic emulsions, using information he had gathered from England. By 1879, he had invented his own emulsion coating machine and in 1880, he began mass producing plates in a small rented loft at Rochester in New York.

The man's name was George Eastman and the business he started was destined to become one of the great names in photography today. The company which Eastman started with Henry Strong was known as the Eastman Dry Plate Company. But although the famous name wouldn't be dreamt up for some years to come, that business, started there in a rented loft, was the beginning of Kodak. Of which, more later.

Let's turn our attention to the cameras of the day.

With the new freedom offered by dry plates, camera styles quickly took off in a number of different directions, but they fall mainly into two categories: field cameras, mostly large, a little cumbersome and rarely used without a tripod; and hand cameras which, as the name implies, could be hand-held.

Classic design

Let's look first at an early field camera that became a classic design on which many manufacturers were to base their ideas. It came from the mind of George Hare, an early associate of Ottewill.

In 1882, Hare designed a bellows camera with the back hinged to the baseboard and the lens board fitted to a movable panel within that baseboard, allowing the lens to be set anywhere along its path. Rack and pinion control gave precise focusing and the camera could be folded after use by sliding the lens panel into the back and then folding the baseboard up to meet it.

A variation on Hare's theme produced the tailboard camera. In this, the front panel was hinged in a fixed position to the baseboard and the back fixed to any one of a number of positions in a panel that again slid within that baseboard. For folding, the back was brought

Three variations on the classic field camera theme. Top: the front panel is movable, the back is hinged to the baseboard. In the tailboard design, below, the opposite holds true - the front is hinged and the back is movable. Bottom: both the front and back panels are movable.

The Lancaster Instantograph

A perfect example of a typical field camera was the Lancaster Instantograph of 1886. Like George Hare's design of four years previous, the back was hinged to the base which folded down to allow the lens panel to travel along its length. By 1888, the Instantograph was offered in four sizes from $4^1/4$ x $3^1/4$ in. up to 12 x 10 in.

The Instantograph was a square bellows camera. Like all early models, the bellows were four-sided and folded with 90 degree corners which put considerable strain on the leather, often resulting in tiny holes. But in the years that followed, a new method of folding became popular. Corners were champfered to make bellows that were actually eight-sided, four of which were extremely short and providing corners with two 45 degree folds, rather than one of 90 degrees.

up flush against the rear of the lens panel and the base, hinged at the appropriate point, folded to meet it, thus protecting the ground-glass screen.

The third variation on the folding design employed movement of both front and back of the camera and was folded after use by dropping the front into the camera's base and then bringing the whole assembly up to meet the back.

If the lens board could be extended on its movable panel beyond the length of the baseboard, the camera was said to have a double extension; movement of the back beyond the baseboard as well gave the camera triple extension.

Early shutters were in the form of accessories extra to the basic camera.

A vital accessory

Before leaving field cameras and venturing into the various types of hand camera, it might be advisable to pause at this point and consider one of the accessories that was becoming more and more popular around this time. While it is not the intention of this book to look at camera accessories in too much detail, this particular piece of equipment can hardly be ignored. Because what started off as an accessory quickly became a necessity. We're talking about the shutter.

As plates increased in speed, it became obvious that a more accurate method of making exposures was going to be needed. No longer was it possible to time an exposure by removing the lens cap, counting thee and replacing it. Something far more precise was needed.

The first shutters were completely independent of the camera. They were bought separately and simply fixed to the front of the lens. In an early design, a box contained a blind that ran over a roller. As the photographer pulled a string attached to the end of the blind, it lifted to allow light through the lens, while the opposite end of the blind travelled down the other side and closed the gap again. The faster the string was pulled, the briefer the exposure.

Another design opened and closed a flap for whatever time was required, the timing again judged by the photographer. From there, it was a simple step to make

The Waterbury 5 x 8-inch Field View Camera from about 1885. For such a camera, the shutter would have been considered as an accessory. This one works by a rubber band.

the flap open by a pneumatic cylinder operated by squeezing a rubber bulb. (It's rather quaint to realize that even now the 'open shutter' setting on modern cameras is still represented by the letter B which originally stood for this bulb.)

Soon shutters began to get more mechanical. A see-saw arrangement of two holes passing the lens was one type and a rotating disc was another. More often than not, these were powered by rubber bands.

Not that all shutters were successful. Some even defeated their own object, operating so violently that they induced camera shake during the actual exposure.

Mousetrap shutters

By far the most popular type of shutter found around the time of early dry plate cameras was one that came to be known as the mousetrap; a roller blind type made, among others, by Thornton Pickard. The principle was as simple as it was effective. A flat wooden box held a fabric blind with a gap in it suspended between two rollers. The rollers were sprung and could be tensioned to give an exposure which was usually anything between 1/15 and 1/90 sec.

The shutter was set by pulling a chord which tensioned the spring. It was released by a lever which could be worked by hand or rubber bulb. Mousetrap shutters came in two different types; one that fitted snugly over the front of the lens and another that fixed to the camera and allowed the lens to be screwed into its front.

The focal plane shutter arrived in the mid-1890s and, like the others, it was first

The shutter is mounted on the front of the lens in this Anthony Patent Bijou camera, made around 1884. This was the smallest of the Anthony cameras.

The addition of viewfinders and focusing scales freed cameras from the tripod at this time. This example is the Bo Peep Camera. with reversible finder, red bellows wood front and interior.

designed as a separate accessory, but one that the manufacturer tailored to fit individual cameras on a slightly more permanent basis. The shutter took the form of a roller blind with a wide slit in it. The photographer had two adjustments to play with; one controlled the speed with which the slit travelled across the plate, the other varied the width of the slit itself. The blind also had provision for a plate-sized gap that was brought into play for focusing.

Then came mechanical and metal between-the-lens shutters such as those produced by Bausch and Lomb in 1891. Exposure on these was governed by metal blades that opened and closed, usually at settings from around 1 to 1/100 sec. In some cases, one set of blades controlled both shutter speed and aperture. It was these mechanical between-the-lens shutters that were usually found on hand cameras. Which leads us very neatly back to our second category of early dry plate camera.

Viewfinders and focusing screens

The basic idea of a hand camera was to take the instrument away from the tripod and make it possible to use it in the hand, and that meant finding a way of using the camera without a focusing screen. So the camera would need first a viewfinder, and second a focusing scale of some sort. Many makers seemed to believe that the addition of these two refinements on a model that was smaller than the average field camera automatically turned it into a hand camera. They were, to say the least, optimistic.

Plates were getting faster all the time, but in many cases they still weren't fast enough for hand-held exposures, and a tripod, while not a necessity, was still a distinct advantage. Which is why we more often hear these early examples referred to as hand and stand cameras. They could be used either way, but there's no prizes for guessing which produced the better results.

So the first hand cameras were nothing more than smaller versions of their field camera cousins, with the addition of a viewfinder and a focusing scale.

Magazine cameras

But as plates grew faster and more reliable, cameras did become truly portable and that was when numerous styles of hand cameras began to appear. There were magazine cameras that held perhaps a dozen plates all loaded and ready for use inside the camera, and in these, plates were changed after each exposure by all sorts of clever and ingenious methods.

Fallowfield's Facile Camera of 1887 is a good example. In this, plates were loaded into a box on top of the camera and, after exposure, were moved to a corresponding box in the base.

Magazine cameras, as we shall see later, were extensively developed and improved upon during the 1890s in an attempt to challenge the growing interest in roll film.

But perhaps more than anything else, this was the time of the miniature, a time when cameras not only became smaller, but took on many weird and wonderful shapes that a few years previously would never have been associated with cameras at all. It was a time when detectives were coming into vogue. In fiction, Sherlock Holmes was beginning to show his face, in America, the famous Pinkerton Agency was making news. Detective was the in word and it wasn't long before it became applied to cameras.

In general, detective cameras were either models small and convenient enough to be hidden at the moment a picture was being taken or cameras that were quite blatantly styled to look like something else. This last type, while known by the overall title of detective

A box-type magazine camera from Lumiere around 1892.

cameras, should more properly be termed concealed cameras.

A good example of a straightforward detective camera that was not necessarily a concealed camera, was Schmid's Patent Detective Camera which, in 1884, introduced Americans to the style. It was in the shape of a small box 6 x 4^1/4 x 8^1/2 in. glancing at a ground-glass screen gave photographers the freedom to take a picture without using something that spelt CAMERA in large letters to onlookers who could then be easily photographed without their knowledge.

In England, there was a model made by the London firm of Watson and Sons in 1886. This was basically a quarter plate bellows camera fitted into a leather-covered box equipped with two viewfinders - one on the top and another on the side. Inside the box and at the back of the camera there was a space large enough to hold some dark slides loaded with plates.

Most small or unobtrusive cameras of the time were referred to as detective

The Scovill Detective Camera, made around 1886.

cameras, but surely the most interesting models were the concealed cameras. As each manufacturer tried to outdo his rival, cameras arrived on the market disguised as bowler hats, shoes, watches, books, walking sticks, tie-pins, shirt buttons, neck-ties, revolvers, binoculars, handbags, drinking cups, Gladstone bags and purses.

The Stirn

One of the best known styles of concealed camera was seen in Gray's Vest camera of 1885, which hid behind a false shirt front. With the false shirt discarded and the rights to the camera in different hands, it became the Stirn in 1886. The camera was designed to be worn chest-high, suspended from a cord around the photographer's neck. He would wear it under his shirt or waistcoat with the lens poking out through a button-hole. Exposures were made by pulling a small ring at the bottom of the device and then, as a central knob was turned, the next plate was brought into position and the shutter tensioned all in one movement. It took six exposures on special circular plates.

Bloch's Photo-Cravate (or Necktie Camera) came from France in 1890. This was a long, oblong, device hidden in a cravat that the suspicious person might have noticed having a lens sticking through where a pin might normally have been found. A small, round knob protruding from the bottom was attached to a cycle chain which brought one of six plates into position for each expo-

The Demon

The Demon was a cheap and cheerful detective camera made in England in 1889 for an English company called the American Camera Company. Which is why the camera is sometimes thought to have been American. It held only a single dry plate and once the picture had been taken, it needed a darkroom or changing bag before the next could be considered; a fact that the manufacturers glossed over in advertisements of the time:-

'Defies detection and can be secreted under vest, in the watch pocket or concealed in the glove. No movement is too rapid for it - the racehorse at greatest speed, the flight of birds, even the lightning flash itself.'

Cameras in disguise...

A few of the cameras from this era that looked like anything but cameras.

Right: The Photo Revolver.

Below: The French Necktie Camera or Photo-Cravate.

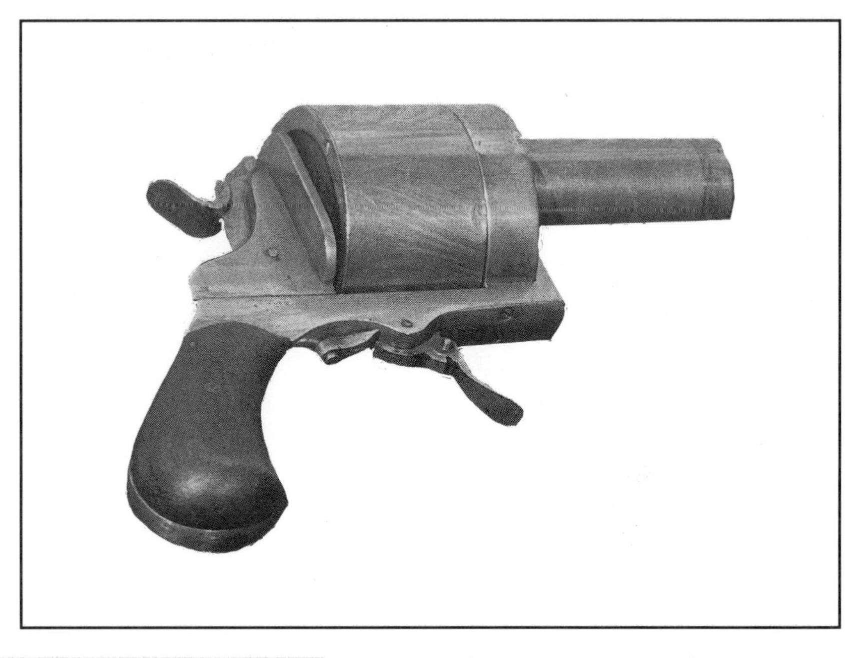

Two watch type cameras. Below, Lancaster's version and, opposite page, the Ticka, showing how the style survived into the early 1900s.

Above: The Stirn, designed to be hung around the neck with the lens poking through a waistcoat button-hole.

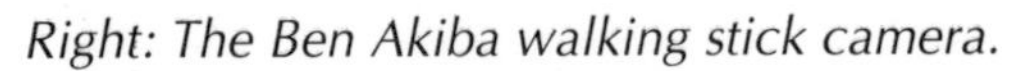

Right: The Ben Akiba walking stick camera.

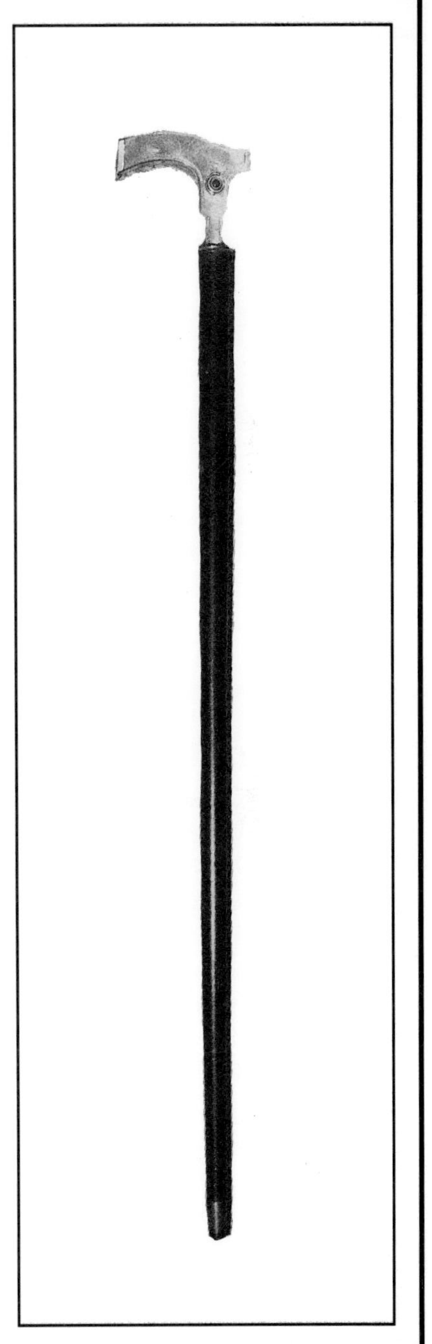

sure. The exposure itself was carried out by means of a pneumatic bulb secreted in the trousers pocket.

The Hat Camera was the speciality of the Adams Company among others and is thought to have fitted into the top of any standard bowler. It was designed to be used by removing the hat and aiming the camera, holding the brim close against the chest. Unfortunately, although designs for the camera have frequently been published, few will lay claim to having seen one.

Then there was the Krugener Patent Book Camera of 1887. This was actually bound in morocco to look like a typical novel of the day. The lens that looked out through the spine was an f/16 meniscus type and the camera had a single speed shutter operated by a string.

Or how about Thomson's Revolver Camera made in 1862. It held tiny plates in the bullet chamber, had a lens down what might have been thought of as the gun's the barrel. It must be said, however, that there was no strong resemblance here to a real gun.

But one of the most popular styles of concealed camera was that disguised as a watch. The Lancaster Company introduced theirs in 1887. When closed, the camera looked like a pocket watch of the day; but pressing a spring on the side shot a series of telescopic tubes into position, much like opening a tiny bellows camera.

According to Lancaster's advert: 'The lens is a very rapid one and can be adjusted for taking portraits, groups or views etc.' The advert, however, didn't explain how.

From plates to film

The watch camera was one of the few designs of concealed camera that made the transition from plates to film, and although we are not dealing with film as such in this chapter, no history of the concealed camera would be complete without mention of one of the most popular. The appropriately named Ticka was made by Houghton in 1906 and it was a watch camera with the lens concealed in the 'winding mechanism'. Yet the really revolutionary thing about the Ticka was not its shape, but its film; 16mm wide, giving a 22 x 16mm image, backed with paper and fed from one cardboard cassette to another. It could almost have been the blueprint for later 110 cameras.

Other watch-type cameras included the Magic Photoret of 1893, The Expo Watch Camera of 1905 and even as late as 1949, a wristwatch version called the Steineck ABC.

The latter part of the 19th Century was, without a doubt, a time for camera innovation. Field cameras, hand cameras, miniature cameras, detective cameras... there was something to suit everyone and every purpose. It seems impossible that any other major design could possibly have been squeezed into so short a space of time. And yet, before the Century was out, three more important styles of camera, two of which are still with us today, had arrived. In 1861, the first single lens reflex was patented, 1882 saw the arrival of the first twin lens reflex and in 1897, the first colour camera was built.

In the 1861 design by Thomas Sutton, a mirror reflected the lens's image onto a screen in the top of the camera. To expose a plate in the back of the camera, the mirror was flipped out of the light path by turning a knob on the side and this action also acted as a crude form of shutter. But although Sutton designed and patented his idea in 1861, evidence points to the fact that the camera was never built. Probably the earliest single lens reflex to actually appear on the market was one made by E.W. Smith of New York and patented in 1884. Other single lens reflexes of the time included names like the Vanneck, Gambier and Miral - all from the 1890s.

The first twin lens reflex

Marion's Academy camera of 1881 is usually recognised as the first twin lens reflex. The camera featured two lenses - one for taking, the other for viewing -

Marion's No.1 Academy. Unlike many that followed it, it had an eye-level, rather than a waist-level viewfinder, but it was, nevertheless, the first twin lens reflex.

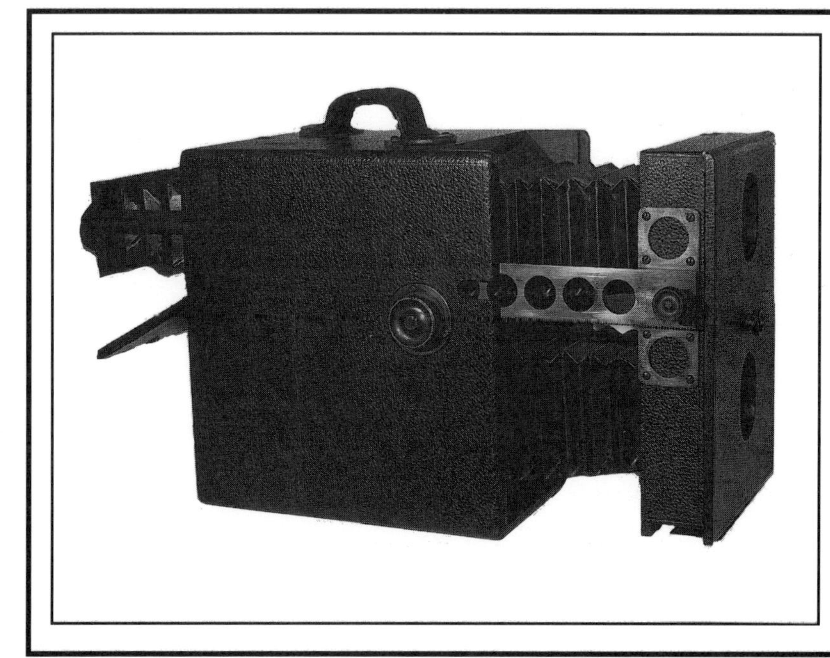

The Graphic Twin Lens Special, an early twin lens reflex. The bottom lens took the picture, while the top lens focused on a ground-glass screen at the rear. Both lenses moved together in the front panel for focusing.

mounted one above the other. The viewing lens projected its image onto a ground-glass screen at the back of the camera immediately above the plate. On the base of the camera, a small magazine held twelve plates that could be moved along by means of a rack and pinion. When the first plate was in position directly below the plane of focus, the camera was inverted so that the plate fell into its rightful place. Then it was dropped back into the magazine after exposure and the magazine moved up another notch for the next plate.

The camera was finished in polished wood with an f/5, 50mm lens and rotary shutter. Like so many of its contemporaries, it summed up the precision and sheer inventiveness of the age.

Black and white into colour

Cameras designed for taking colour pictures date from 1874 when Louis Ducos du Hauron patented the Chromographoscope, built eventually in 1897. But although this can be considered as probably the first colour camera, it was by no means responsible for producing the first colour photograph.

Colour photography had been tried and even proved possible right back at the start when Niépce produced that first heliotype in 1826. There is evidence to show that he and Daguerre actually succeeded in taking colour photographs of a kind, though they never discovered a way to fix them and so no substantial proof of their claims exists today. Neither is it clear what methods they used.

In 1839, Herschel actually obtained a coloured photograph of the spectrum with which he forecast the possibility of serious colour photography; but again there was no way of fixing his image.

The man who eventually set colour photography on the right track was an Englishman named James Clerk Maxwell. In 1861, at London's Royal Institute, he showed for the first time how the human eye was sensitive to only three colours - red, blue and green - and that all other colours were no more than combinations of these three primaries.

Maxwell had three black and white lantern slides made of the same subject, a tartan ribbon. Each plate was taken through a filter of one of the primary colours. The plates were projected, each with their appropriate filter in front of the lens. The three images were superimposed on a screen and the result was a colour picture.

The way was now clear for colour cameras to be made. What was needed was a single camera that could take three exposures on three different plates through three different filters - all at the same time. The Chromographascope was the first. In this, a system of straight mirrors and semi-reflecting mirrors allowed the lens's image to be reflected around the camera onto three different plates, each through one of the primary colours. The camera was also used as a viewer when the developed plates were put back in their appropriate places.

This was followed by similar cameras such as the Melanochromoscope of 1899; and models such as that made by Lancaster for E.T. Butler in 1905 which used a Compound shutter, f/6.3 lens and bellows for focusing.

Colour cameras were many and various, but all relied on the one basic precept: they were all methods of producing three different images of the subject. It was another few years before the inventors realized that they were on the wrong track. What was really needed was not so much a special camera, as a special plate - one that could be used in any ordinary camera of the day to make a coloured image. And that wasn't as far off as might be expected.

The first years of the 20th. Century also saw another innovation that would actually be echoed more than 80 years later. It came from an American called Frederick Eugene Ives who, in 1903, produced what he called parallax stereograms. In these, a stereo pair of of pictures was printed as a series of narrow, alternating strips. A special grid was then placed at a specific distance from the print, so that each alternate strip was blocked from one eye. The right eye, therefore, saw one side of the stereo pair, the left eye the other. The result was stereo photography without any further viewing apparatus. It was a technique that would be repeated several times during the coming years, culminating in the Nimslo system of the 1980s, to which we will be returning towards the end of this book.

6. You press the button, we do the rest

TOWARDS THE END of the 19th Century, one man did more to popularise photography and cameras for the man in the street than anyone else before him. His name was George Eastman.

We've already seen Eastman as one of the early pioneers in gelatine dry plate manufacture, but there was a great deal more to the man than that. In the years that followed the start of his dry plate business, he turned his attention to the possibilities of roll film and in 1885, he introduced the Eastman-Walker roller slide, a device that clipped onto the back of a conventional plate camera in place of the plate holder and held a twenty-four exposure roll of paper film. After exposure, the emulsion-coated paper was developed and then oiled to make it transparent for printing.

A year later, in 1886, Eastman began making his roller slides for use with a new type of paper-based film called stripping film and although he wasn't the first to use it, he was the first to use it extensively; first in his roller slide and later in a camera.

Eastman's own brand of stripping film consisted of a layer of soluble gelatine laid down on a paper base, followed by a layer of collodion and then a sensitized emulsion. After exposure, the film was developed and fixed in the usual way, then attached to a glass plate coated with glycerine. Hot water was then poured over the paper, the soluble gelatine melted and the paper stripped off, leaving the image behind on the glass. Finally, the now reversed image was transferred to a damp sheet of gelatine, from which the final print was made in the usual way.

Messy and complicated

The process was both messy and complicated, and yet it was with this that Eastman launched his first camera. His aim was a model which would bring photography within easy reach of everyone, a simple camera that anyone could use anywhere at any time - given, of course, the right lighting conditions.

Eastman launched his camera with a previously unheard of name that was destined to live on well after the camera and long after Eastman himself. His idea was to find a truly distinctive trademark. He wanted a short, sharp word that could be easily

The first Kodak

Left: The first Kodak camera, launched in 1888 by George Eastman.

Below: One of the circular pictures taken by the Kodak.

The original Kodak was a detective camera, a box design about 7 in. long that took 100 exposures at one loading. It had a fixed focus lens and a fixed shutter speed of around 1/25 sec - fast enough to allow hand-held exposures, slow enough to register on the slow film. It took circular negatives with a $2\frac{1}{2}$ in. diameter that utilized the whole of the lens's projected image.

But the thing that appealed to Mr. Average was the camera's extreme simplicity. Eastman launched his new era in photography with the slogan, *You press the button, we do the rest* and that really summed up the whole process. The camera sold for $25 (about £10 at that time) which included an already loaded film. Taking a picture was easy. A string cocked the shutter, a button released it and a key on the top wound on the film.

When the 100 exposures were finished, the photographer sent the whole camera back to the works where it was unloaded and the film developed and printed. The first roll of film was processed free and the enclosure of $10 (£4) guaranteed a new roll of processed-paid film loaded in the sealed camera on its return.

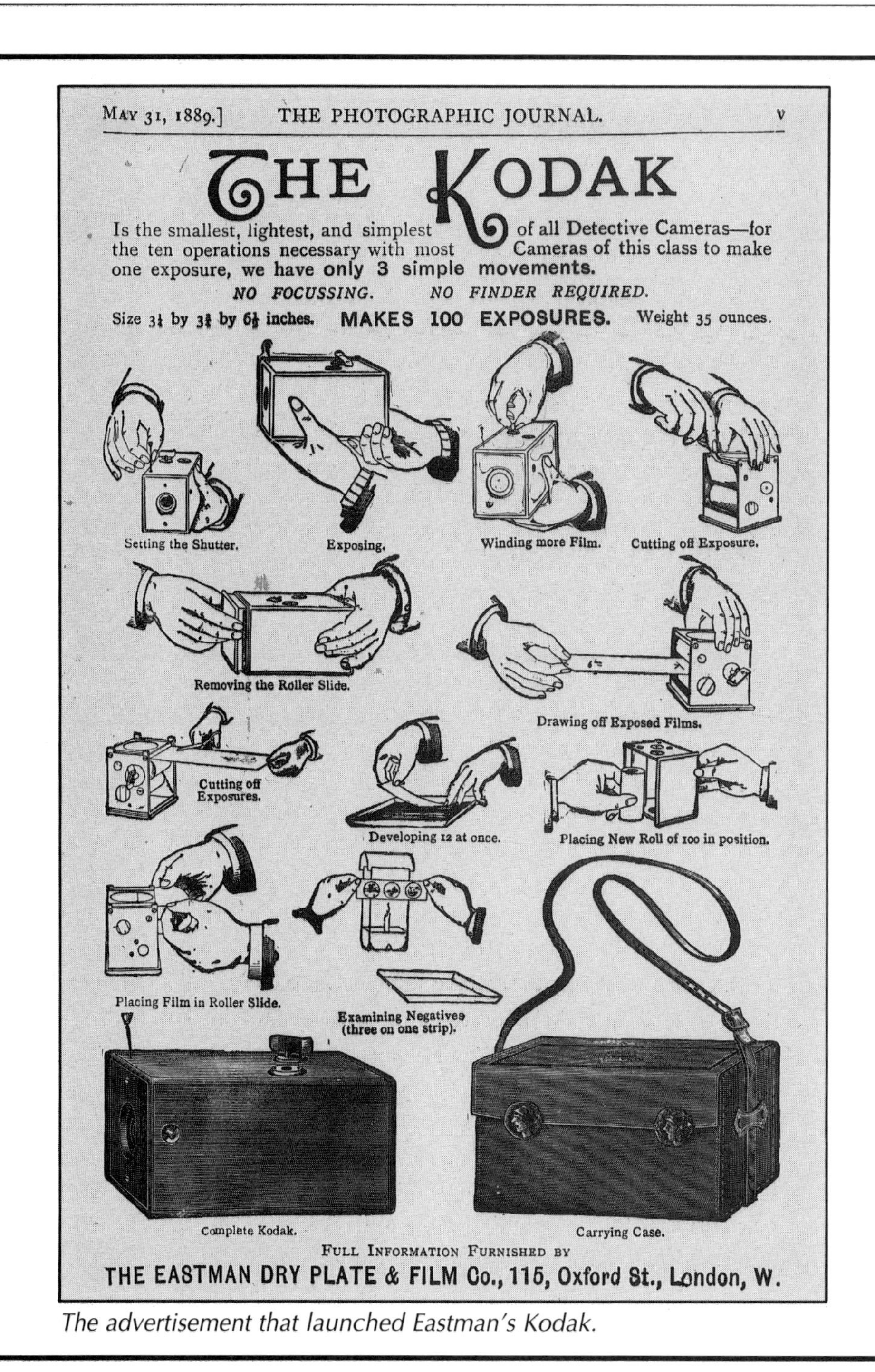

The advertisement that launched Eastman's Kodak.

pronounced, had no current meaning in any language and was difficult to miss-spell. He eventually came up with a completely new word that was as unusual as it was distinctive and with this, in 1888, he launched his camera. He called it simply, The Kodak, a name never heard before that time.

In the same year as The Kodak was launched, an English photographer, then living in America, introduced emulsion-coated celluloid in the form of cut film of various sizes. And eighteen months before that, the Rev. Hannibal Goodwin had applied for a patent on a similar idea applied to flexible roll film. The patent, however, was not granted at that time. It fell to Eastman to actually make the idea viable and in 1889, The Kodak was sold with a celluloid roll film in place of the less practical stripping film.

The fact that Eastman's company had actually won a patent a few years after Goodwin's application led to a law suit which, after twelve years in the courts, ended in Eastman having to pay out the sum of $5,000,000. As the price for a world monopoly on the most practical method ever of taking photographs, Eastman must have considered the sum worthwhile. Transparent flexible film began a new era in photography and one that has not been susperseded even today.

Eastman's rivals

But although Eastman was always the great innovator, he was by no means the only one in the field. It was a time for eager competition between camera manufacturers and the fact that Eastman held the rights to make celluloid roll film did nothing to deter rival companies from coming up with new camera designs of their own. In the same year as The Kodak switched to celluloid film, the Boston Camera Company introduced their Hawk-Eye Camera, a box design that featured adjustable bellows within its body and which gave the user the best of all possible worlds in its choice of sensitive material: single exposures could be made on 5 x 4 in. dry plates or cut film and 100 exposures could be made at one loading on roll film.

Not that all new designs took roll film. There were still plenty of plate cameras about and new designs were appearing every day. Jumelle cameras were a popular style of the 1890s, They took their name from the fact that their distinctive style was that of a squat pyramid, rather like a pair of binoculars enclosed in a case - and the French for binocular is jumelle.

The first Photo-Jumelle was built by a scientific instrument maker named Jule Carpentier. It took the form of a pair of binoculars with one lens for taking the picture and the other acting as viewfinder. The camera was commercially produced in 1892 and took twelve 2½ x 1¾ in. dry plates in a magazine.

Although the name jumelle was originally applied to twin lens cameras of this style, it eventually came to refer to any camera of a tapering design that looked vaguely like a pair of binoculars. Such a camera was the Photo-Etui-Jumelle of 1893, which took single exposures on 4³/4 x 3¹/2 in. plates or the Steno-Jumelle of 1894 that held eighteen plates in a magazine and had a direct vision viewfinder mounted on top.

The late 1890s and early 1900s saw a tremendous rise in the popularity of stereo cameras as manufacturers everywhere began producing stereo counterparts of their standard models. The jumelle type of course made the most obvious choice for such adaptation, being already based on opera glasses and therefore the perfect design for a two lens camera.The Bazin et Leroy Stereocycle of 1898 was a good example. It actually looked like a pair of rather square binoculars. Not so the Bellieni Jumelle being made about the same time. This was a box form camera with a magazine back to hold thirty-six 3¹/2 x 3 in. plates. As well as a rising and crossing front, the camera sported six-speed rotary shutters on each of the f/8 lenses.

Kodak's second generation

Meanwhile, George Eastman, never a man to rest on his laurels, was pressing on in new directions. After his first Kodak, he introduced the No.2 (only then did his original camera become known as the No. 1). This second generation Kodak took the larger picture size of 3¹/2 in. Then, in 1890, he introduced the first Kodak Folding

The 'A' Ordinary Kodak Camera from the 1890s. It had a special sector shutter and took 24 exposures 2³/4 x 3³/4 in.

Kodak cameras took several different paths around the turn of the Century. This example, from 1897, is the No. 4 Folding Kodet.

Cameras: the No.3, No. 3 Junior, No. 4 and No. 4 Junior. These were in the style of a box with one side that dropped down to support a lens that slid out of the box on bellows.

Like the first models, the film for these folding models was loaded and processed at the factory. But then, in 1891, Eastman's company marketed a daylight-loading camera, together with roll film that could be loaded or unloaded for the first time by the customer. The new cameras were larger and had larger lenses that for the first time on a Kodak camera necessitated the use of a focusing scale.

Of course, while Eastman held the rights on roll film, he enjoyed the best of both worlds. Every new roll film camera designed by a rival company merely meant more sales for his film. But although roll film presented possibly the most significant advance to date, it was by no means the cheapest and, compared with plates, the quality still left a lot to be desired.

Cut film camera

The Frena camera introduced by R. and J. Beck in 1892 used neither roll film nor plates. It took celluloid cut film in packs of twenty or forty, arranged like a pack of cards with a piece of opaque card between each sheet. When the pack was loaded

The Kombi

The Kombi appeared in 1892. It was a tiny box camera only 2 x 1 1/2 x 1 1/2 in. that took twenty-five exposures at a time on specially spooled roll film only 1 in. wide. But the feature that made the Kombi so unusual was the fact that it could also be used as a viewer. A panel was removed from the camera's back and the pictures, wound through on the film rollers, could be viewed through the lens.

into the camera, a mechanism within matched up to alternate notches in the film sheets, allowing each to drop into a chamber on the back of the camera after exposure, while holding the rest ready for the following exposures.

But despite its drawbacks, roll film was catching on and being adapted to a number of different types of camera, such as the Kombi or one unusual model from France, which used roll film in a very different way. The Cyclographe was a panoramic camera that revolved on a clockwork motor through anything up to 360 degrees, while the film moved in the opposite direction. The result was a negative 3 1/2 in. wide and up to 30 in. long.

Kodak's red window

The following year, in 1895, Eastman introduced another new model from his works. The pocket Kodak was a box camera, but one that featured an idea that to many seemed revolutionary. For the first time in a Kodak camera, the film was backed with paper which was numbered. The numbers could be read through a small red window in the camera's back as the film was wound from one exposure to the next. This new idea was known as cartridge film.

Such is Eastman's reputation for inventiveness that he is often credited with being the first to introduce the red window idea; but in fact he was using it under licence at that time from the Boston Camera Manufacturing Company who introduced the idea three years earlier in a box camera of their own called the Bulls-Eye. In the end, the red window became such an essential part of the camera design that Eastman bought out his rivals rather than pay them royalties on every camera he produced. He then carried on making the Bulls-Eye under the Kodak name.

Roll film was being used and adapted to many different ends and yet, strange as it may seem, 1895 also saw the birth of one of the most famous and versatile plate

The Sanderson

Sanderson's innovation was to take a basic field camera design and modify it so that the lens panel was supported in four slotted arms - two on each side of the body pivoted on the extending panel within the base.

The four arms were made to move independently and the lens designed to slide up and down them, to be locked in any position along that movement, including out of square to the plate if necessary. The result was a camera with a full range of rise and fall, swing and tilt; all controlled by two knobs on the side of the lens panel, while the photographer watched the result on the ground-glass screen.

Later models were fitted with a tallbody feature which meant the back of the camera was made slightly taller than usual to accommodate the bellows when the rising front was at the top of its throw.

cameras ever produced. We're talking about the Sanderson. Its designer, Frederick Sanderson, was a cabinet maker and stone mason. He was interested in both photography and architecture. It was when he tried to combine his two interests, using cameras of the day, that he ran into trouble. Converging verticals and distortion may actually enhance a pictorial worker's picture, and of course, neither is a problem to the portraitist. But if, like Sanderson, you wanted to make an accurate record of a building, then you needed something more than a straightforward field camera. A rising front of course helped, and crude versions of these were around in Sanderson's day, but it by no means solved all his problems.

So, like many others, Sanderson set out to design a camera for his own personal

Plate camera manufacturers tried several tricks to compete with roll film. One was the falling plate camera that could be pre-loaded with plates, changed between exposures at the press of a catch on the top of the camera.

needs and found he had invented a classic that was to become a historic milestone in the history of the camera.

The plate camera was a long way from being dead and many manufacturers still maintained that plates would win out in the end. Even so, roll film did give photographers a number of exposures on a single loading and that, it had to be admitted, was an advantage. So rather than build cameras to take Eastman's film, the plate manufacturers came up with a compromise. The result was the falling plate camera, a very simple magazine style that became immensely popular in the 1890s.

Falling plates

The camera was a box design. A number of plates were held in plate holders stacked in the back of the camera one behind the other and held tightly together by a large spring in the camera's back. Moving a knob on the top of the camera operated a catch that allowed the front plate to fall into the camera's base while holding the next in position for the following exposure.

Shutters were simple and single-speeded as a rule, and focusing, when included in the design, was often by a series of supplementary lenses that slid into position in front of the taking lens. Falling plate cameras could be bought for as little as £3 in the 1890s and by 1900, manufacturers had their prices down to as little as £1.

Compared to Eastman's Kodaks, they were cheap, but Eastman knew when he was onto a winning streak and, ignoring the opposition, he pushed ahead with new designs of his own. In 1898, he introduced the Folding Pocket Kodak, the first camera with an all-metal case. It took $3\frac{1}{4}$ x $2\frac{1}{4}$ in. negatives and the lens panel was supported on hinged arms that folded back into the body, making a flat box $6\frac{1}{2}$ in. long by $1\frac{1}{2}$ in. thick which was easily slipped into a pocket. The shutter had two separate controls for time and instantaneous exposures.

The all metal Presto.

Two more models of the same basic design were introduced in 1899, the year that Kodak also came up with a panoramic camera. The Kodak Panoram took five photographs on a roll of film, each one 8.75 x 19.5cm. A touch of the shutter release caused the lens to swing through 142 degrees, exposing the film as it went.

Gravity shutter

The same year saw the introduction of an unusual American miniature called the Presto which could be used with $1\frac{1}{4}$ in. plates or roll film. An ellipse-shaped metal box in design, it had a gravity shutter that was set by simply turning the camera upside down.

But how ever much we stray to other makers, we must return time and time again during these years to George Eastman. In 1900, a No. 3 version of the Folding Pocket Kodak appeared and took quarter-plate roll film. It was this

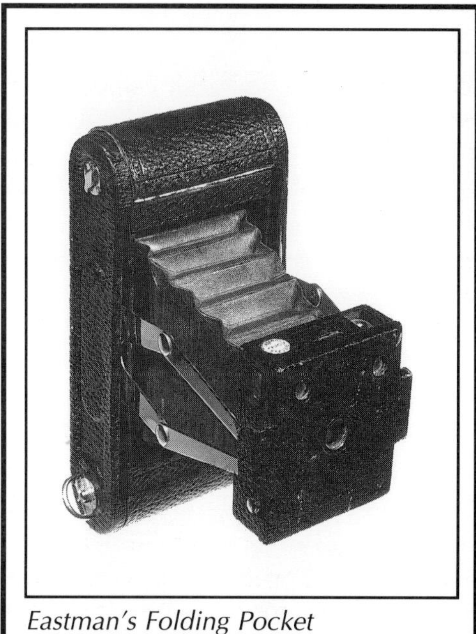

Eastman's Folding Pocket Kodak of 1898.

model that took on the shape with which folding cameras would be associated for another half century or more. Pioneered by Kodak, copied by so many others, the design took the original flat box and added a hinged baseboard one-third the way

The Mammoth

The officials of the Chicago and Alton Railroad Company wanted a picture of their newest luxury train for the Paris Exposition. They wanted the picture to be impressive, perfect in every detail, but, most of all, they wanted it to be BIG. So they commissioned a special camera to be built for the purpose. What they got was a monster called the Mammoth,.

The Mammoth, 13 ft. long and operated by a team of fifteen men, weighed 1,400 lbs. It was moved about the country on a specially made railway truck. Developing the 500 lb plate took 10 gallons of chemicals to produce an 8 ft. x 4 ft. 6 in. image.

The judges at the Paris Exposition were so overwhelmed that they awarded the subsequent picture of the train Grand Prize of the World. After which, having done its job, the Mammoth was quietly dismantled and never seen again.

along its front. When folded down, the lens panel was pulled out along it and, although the camera could easily be used in the horizontal position, it is in the vertical position, with the baseboard supported on its small metal leg that we most often think of the style.

So the new century began. Camera making became a highly mechanised business. Metal took the place of wood, and shutters, already recognised as necessities rather than luxuries, became more sophisticated with each new model that appeared.

With the start of the Century, came a new name that moved straight into the history books as 1900 saw the arrival of the first Brownie, a box camera aimed at bringing photography to the up and coming generation of boys and girls who previously had only ever seen dad using a camera. The name was chosen deliberately to appeal to the youngsters and the cardboard box in which this basically simple box camera was sold decorated with pictures of brownie characters.

How the Kodak Brownie was advertised.

The early part of the 20th Century was a time when the first amateur enlargers began appearing on the scene, but at the very beginning of the Century, the only real way to get a big image was to use a big plate, and in America in 1900, the idea was taken to almost ridiculous lengths with the Mammoth, undoubtedly the world's largest camera

Practical colour process

In 1903, the first practical method of colour photography was patented by Louis and Auguste Lumière. It was called the Autochrome process. While we are dealing more with cameras than processes in this book, it might be as well to pause here to look at this particular method since it enabled photographers for the first time to take colour photographs with their everyday conventional cameras, rather than using the special separation models that had been needed up until then. The process was put on the market in 1907.

The Autochrome process was the most positive step that had yet been taken in

Autochrome process

In the Autochrome process, minute grains of potato starch were first dyed, each one orange, green or violet. These were then sorted together so that each of the three colours was present in an equal proportion and then scattered over a glass plate, nearly 10,000 particles to the square millimetre. When pressed, they then formed a lamination of tiny coloured filters. The gaps between the grains were dusted with charcoal so that light could penetrate the plate only through the grains and the whole thing was then covered with a transparent varnish.

This was then coated with a panchromatic emulsion and the plate slotted into the camera, glass side towards the lens, so that the starch grains were acting as primary coloured filters in front of the emulsion. A yellow filter was placed over the lens and the exposure made.

The plate was developed in the normal way and then further developed to make a positive. With the starch grains still intact, the plate was then held up to the light or placed in a viewer, when a true representation of colours would be seen

colour photography and it proved tremendously popular. By 1913, the Lumière factories were making up to 6,000 Autochrome plates a day and production actually continued until as late as 1932 when the firm adapted their process from plates to roll film. The result was Lumicolor.

Back to Kodak. In 1904, the first folding Brownie appeared, a horizontally styled folding camera, this time with a baseboard that dropped down from the centre of a flat oblong box. The folding Brownies, made after a year in two sizes, were distinguished by their red bellows and polished wood lens mounts.

Panoramic pictures

In 1904, a new panoramic camera called the Cirkut arrived from a division of Kodak. Its principle was totally different to their earlier Panoram. Mounted on its own specially made tripod, the Cirkut was capable of revolving a full 360 degrees by means of one of a number of cogs in the camera's base that matched up to a huge cog on top of the tripod. Several different roll film sizes were

used, each mounted in a magazine fitted to the camera's back. As the camera revolved one way, the film moved in the opposite direction across a vertical slit through which the exposure was made.

A lever on the side of the magazine controlled the speed or rotation and movement and, with no shutter as such, this movement also governed exposure: the faster the rotation, the less exposure given; the slower the rotation, the more exposure.

A Kodak Cirkut, together with one of its panoramic pictures.

The Cirkut was made right up until 1949 and many are still doing the rounds today, used for taking those famous school photographs that call for a thousand boys to be arranged in a tiered semicircle while the camera, mounted on its tripod in the centre, scans the whole group.

Let's turn our attention once again to shutters, since by now, they were an all important part of camera design. Most camera makers used shutters made by separate specialist manufacturers and the most common design of the period we are discussing was the pneumatic shutter. In this, the exposure was determined by the movement of a tiny piston within a metal cylinder.

In some, this meant that air leaked through a by-pass in the piston, while in others, the rate of movement was controlled by the amount of air allowed to leak out of the cylinder, and often the exposure was represented by the actual speed of the shutter blades' movement.

More common, however, was the method used in more sophisticated pneumatic shutters such as the Unicum where the blades always moved at the same speed.

The No.3 Folding Pocket Kodak, launched in 1900, marked the start of a simple folding design that would remain popular for a half a century or more.

Sophisticated shutters were based on a pneumatic principle to control the length of time for which the blades were open. This is a good example from Unicum.

Exposure relied on the time that they were actually open and that was controlled by the variable length of the piston stroke within the cylinder.

The main fault with these mechanisms was the way working parts were exposed to the air, making them particularly susceptible to dust. The obvious next step was to put a case around the moving parts.

The Compound shutter

Perhaps the best known shutter of the early 20th Century was the Compound. It made its first appearance during the first years of the Century and in its improved and most popular form in 1911. The shutter was finished in black enamel and featured two levers, one for tensioning the shutter and the other for releasing it. The mechanism that worked the blades was totally enclosed and if a small cylinder was to be seen on the top, this was there only to facilitate the use of a pneumatic bulb release.

Up until this point, shutter blades were set alongside another set of blades that

The fifteen-lens Royal Mail Camera.

controlled the aperture; but in the Goerz and Volute shutters, one set of blades did both jobs: set f/8 at 1/25 sec and the blades would only open as far as that aperture during the exposure, set f/6.3 at 1 sec and the blades would open right up to that aperture for a second's exposure...and so on.

To write about shutters in any real detail could easily take up the rest of this book. So while no history of the camera could possibly be complete without a mention of the part they played, there is no room here to do the subject its full justice. Back to cameras.

Despite its rising popularity, roll film was still an expensive luxury compared with plates, and alternatives that gave the best of both worlds were being explored. Cameras such as the Saint-Etienne Universelle of 1908 were made as plate cameras with provision for an easy-to-attach roll film back.

On the novelty front, there was the Royal Mail camera made by W. Butcher and Son in 1908. In some ways, the camera was a throwback to the wet plate days; not in its choice of plates, but in the fact that it took multiple pictures - three, six or twelve depending on the model - through as many lenses on one quarter-plate exposure.

The resulting image fitted the portrait space on a postage stamp and special masks were available to give the finished print the look of a stamp with the chosen subject's face on it. Hence the name Royal Mail Camera.

Miniature breakthrough

But more than anything else, the time was right for a simple miniature, and for once, George Eastman was beaten to it. In 1909, the British firm of Houghton's Ltd. introduced the Ensignette, a collapsible bellows camera with a choice of meniscus or f/5.6 anastigmat lens.

Up until then, most miniature or vest pocket cameras had been made for plates or sheet film - the Goerz Tenax of 1908 for instance - so Houghton's made a big thing about the fact that the Ensignette was 'completely self-contained'. In other words, it took roll film, making a separate film or plate holder unnecessary.

From Houghton's, the Ensignette (right) and Eastman's answer, the Vest Pocket Kodak.

Naturally, Eastman wasn't to be outdone for long and in 1912, he introduced the Vest Pocket Kodak, a folding design that measured no more than $4^{3}/_{4}$ x $2^{1}/_{2}$ x 1 in. when folded and made to take 127 film.

Over the next few years, Eastman introduced two new additions to his basic cameras. The first in 1914, was an idea never tried before and never copied since - the autographic back. These specially designed camera backs were fitted to Folding Pocket and Vest Pocket Kodaks, and featured a small, rectangular trapdoor, together with a metal stylus. As each picture was taken, the photographer could snap open the trapdoor and, providing he was using special autographic film, write details of time, place, exposure, or anything else that appealed to him directly onto the film. When developed and printed in the normal way, the information would then be found written at the bottom of the print. The feature lasted until 1932.

First coupled rangefinder

Eastman's second major addition to his cameras came in 1916 when his company issued their No 3A Autographic Special with a coupled rangefinder. It was the first of its type and remained the only coupled rangefinder camera on the market for the next twelve years.

Innovation from Eastman at this time: the autographic back (above), launched in 1914 and, in 1916, the first camera with a coupled rangefinder, situated below the lens.

There were three other types of camera that grew to fame during the period we are discussing here. One was the strut camera. These were bellows cameras in which the front panel was held rigid at a fixed distance from the film plane by four struts, one on each corner of the camera. Focusing was achieved by movement of the lens on its panel and sometimes by movement of the struts. For carrying, the struts were disengaged and folded into the body, followed by the lens panel. The Ensignette and Goerz Tenax already mentioned were strut cameras, as were many of their larger brothers made by firms like Zeiss and Ernemann.

Lazy tongs

The second style, like the first, also employed struts but the difference lay in the fact that they were criss-crossed like scissors and known as lazy tongs. The idea was pioneered by Arthur Newman who looked at the ordinary folding cameras of the day and decided the mechanisms employed were not good enough to hold the lens panel permanently rigid and parallel to the film plane. Thus was born, the Sibyl range of cameras that were to continue in popularity until well after the First World War in styles for roll film, film packs and plates.

The third style that really came into its own between 1900 and 1920 was the single lens reflex. The most popular design was that followed by cameras like the Auto-Graflex of 1907, the Pre-

Lazy tongs were a form of strut, pioneered to give more rigidity to the front lens panel, while keeping it absolutely parallel to the film. The style was seen on the Sibyl range of cameras.

mograph of 1908 or the Ica L'Artists of 1910. They were essentially large boxes with a front panel that moved for focusing on bellows by means of a rack and pinion, and with a large hood that folded out of the top to shield the focusing screen. Because of the reflex principle involved, shutters were naturally focal plane types.

For every new camera, there was a new design, and yet there was still one design still very much in its embryonic stages during the years immediately prior to the First World War, but one which in due time would completely revolutionize the history of the camera.

In 1913, in Germany, a man named Oskar Barnack designed and built the prototype for a camera capable of taking 35mm cine film. It had an all metal body and collapsible lens with a pivoted cover to block light from entering the camera as the focal plane shutter was rewound. The camera has gone down in photographic history as the prototype Leica.

7. The birth of 35mm

OSKAR BARNACK WAS HEAD of an experimental department in the German optical firm of Leitz, he was a keen amateur photographer and he enjoyed hiking; three factors which contributed to the birth of the Leica.

Tired of carrying a large plate camera with all its attendant paraphernalia on his hiking trips, Barnack was turning his thoughts to the possibility of a versatile miniature as early as 1905. In his attempts to produce a serious professional camera that could be hung round the neck or slipped into a pocket, he began his experiments by cutting down large glass plates and using them in a specially built miniature camera. But on enlargement, the coarse grain gave inferior results that were totally impractical.

Oskar Barnack - father of the Leica.

It was later, while still working for Leitz, but this time on a motion picture camera, that the answer came to him. The 35mm film used in these early movie cameras was obviously designed to produce the maximum sized enlargement from the minimum sized negative - after all, what better enlargement is there than a cinema screen? It was the breakthrough for which Barnack had been searching.

The first Leica

He built a miniature camera for use with 35mm cine film and the result was the prototype Leica. That was in 1913; it was to be another ten years before the idea was further developed. In between came the First World War.

The genesis of the Leica as we know it today lies in this prototype produced by Oskar Barnack in 1913, ten years before the Leica I.

Just before the war, two significant companies were formed on the other side of the world. In 1917, an admiral in the Imperial Japanese Navy named Mitsubishi established a company called Nippon Kogaku (meaning Japanese Optical Company) for the purpose of supplying specialist optical and measuring equipment for military use. In 1919, also in Japan, the Asahi Optical Company was started as an eyeglass manufacturer.

At the time, neither company had any thoughts about cameras. But in time, both were to become famous for two of the best known names in single lens reflexes: Nikon and Asahi Pentax.

The First World War came and went and then, between 1923 and 1924, thirty-one Leica cameras were hand-made and used to test public reaction. They had f/3.5 lenses, the focal plane shutters were calibrated, not in seconds, but in millimetres that represented the width of the slit in the blind and, for the first time in the history of the camera, a double exposure prevention device was built in.

The reaction from photographers and scientists alike was, in some cases, sceptical. But despite the fact that some took an instant dislike to the new miniature, Dr Ernst Leitz, head of the famous optical company, decided to take the risk, even though his company had never before made a still camera for the general public. The name Leica (derived from LEItz CAmera) was registered as a trademark and the camera went into production. The Leica I first appeared at the Leipzig Spring Fair in 1925.

Leica - 1925-1940

Screw-fit Leicas ran from 1925, when the Leica I was introduced, through to the late 1950s, with the IIIg, the last of the type and actually made after the bayonet fit M-series had been introduced. Here is a very brief run-down of the first fifteen years, up to 1940.

■ **LEICA Ia (1925)**: The shutter on this first model was speeded 1/20-1/500 sec. plus Z for time exposures. Very early models had a Leitz Anastigmat lens, followed by cameras with an f/3.5 fixed Elmax, after which the majority featured a fixed Elmar. Built-in coupled rangefinder and slow speeds (1 to 1/8 sec inclusive) were missing. A separate rangefinder could be used. Metal parts were black enamelled or nickel-plated.

The Ib model that arrived in 1926 was also known as the Compur Leica because it used both rim and dial set Compur shutters. The Ic introduced interchangeable lenses in 1930, offering a 35mm, 50mm and 135mm.

Leica Ia

Compur Leica

■ **LEICA II (1932):** For the first time, this model offered a built-in rangefinder coupled with the lens, but in other respects it was similar to the Leica Ic. All metal parts were either black enamelled and nickel-plated, or satin-chrome.

Leica II

■ **LEICA III (1933):** This had a built-in rangefinder, coupled with the lens and, for the first time, slow shutter speeds controlled from a separate knob beside the lens. The new speeds, coming into operation with the main shutter dial set at 1/20, were: 1, 1/2, 1/4, 1/8 sec. The other speeds still ran from 1/20 to 1/500 sec, plus T for very long time exposures. A magnifying system was built into the rangefinder. Metal parts were black enamelled and nickel-plated, or satin-chrome finished.

■ **LEICA IIIa (1935):** This was similar to Model III, but with a shutter speeded up to 1/1000 sec. It was also fitted with carrying eyelets on either side of the camera body. All metal parts were satin-chrome finished.

■ **LEICA IIIb (1938):** This was similar in all other respects to Leica IIIa, but now, for the first time, had the rangefinder and viewfinder eye-pieces in one twin-sight opening.

Leica IIIa

■ **LEICA 250 (1934):** Launched for special purposes of the professional photographer, for times when more than 36 exposures at one loading were desirable. To meet these requirements the camera was supplied with a film chamber which could be loaded with 10 metres of film, permitting 250 exposures to be made with a single loading. The camera was fitted with two spool chambers, one of which served as take-up spool. For the rest, the Leica 250 was similar to the model IIIa.

Leica 250

■ **LEICA IIIc (1940):** Similar to model IIIb, but with a slightly longer body. The camera body with rangefinder housing consisted of one piece of material. The film counter no longer made a whole revolution when winding on film, but moved on by one division.

The shutter was reconstructed and the slow speed knob carried the speeds from 1/30 sec. down to 1 sec. and T. Later models had a locking stud on the slow speed knob to prevent its turning by accident away from its standard setting. Experimentally a few cameras were fitted with a built-in 15-seconds delayed action release.

The Leica story, with its many different models could easily fill a book on its own, so we must skip over it as briefly as is justifiably possible. The panel on the previous page, then, deals with the first fifteen years which launched and established the name, so having seen the Leica on its way, let us return to earlier days.

The Leica, more than any other camera, popularized the start of 35mm photography, but it was by no means the first to use cine film for still pictures. Despite the fact that early movie film gave soot and whitewash effects, cameras were designed for its use almost as soon as the film itself appeared.

The Minigraph of 1915 was a good early example. It carried enough film for fifty exposures, each measuring 18 x 24mm and sported an f/3.5 lens. The wooden-bodied Sico of 1920 had a Compur shutter and took twenty-five 30 x 40 mm exposures on paper-backed unperforated 35mm film. The Esco, made in Nuremberg in 1926 made a great play about the number of exposures it could take at one loading. Thanks to a couple of large round chambers on each side that dwarfed the camera itself, the Esco was capable of taking 400 pictures, each one measuring 18 x 24mm. And there was the Debrie Sept that was both a still camera and a short-run movie camera.

The Minigraph, an early 35mm camera.

New lenses

One of the biggest advances in the 1920s was in lenses. Wider and wider apertures were appearing all the time and in 1924, the Ermanox was introduced, sporting the previously unheard of top aperture of f/2.

The Ermanox was designed for single exposures on $2^1/4$ x $1^3/4$ in. plates and rapidly became popular as the first truly candid camera. It measured only $4^1/2$ x $3^1/2$ x $2^3/4$ in. and was sold with the slogan. What you see, you can photograph. A year before the Leica I, it revolutionized available light photography.

We are dealing here with an age when films and plates went side by side.

The Debrie Sept

Was it a still camera, a cine camera, a projector or an enlarger? It was, in fact, the Debrie Sept, a highly unusual camera from France.

The Debrie Sept was made first in 1922 by Andre Debrie of Paris. It was advertised as a three-in-one camera because not only could it take snapshots and time exposures, but with what can be considered as an early form of motor drive it could also be used as a movie camera.

It took 35mm film loaded into special cassettes and its main drawback was its single shutter speed. Exposure had to be varied by the aperture alone.

But the Sept was more than a camera (or more than three cameras if the advertising was to be believed). A panel opened in the back to allow the attachment of a light which turned the camera into an enlarger or projector.

Slots could also be opened top and bottom to feed through a roll of already developed negatives which then lay against the unexposed film. If the lens was then covered and the rear panel opened, contact printing to make positive pictures could be carried out actually within the camera.

The Ermanox - taking advantage of new lens technology to make low light, hand-held photography a real possibility.

While cameras like the Leica were extolling the virtues of roll film, the smaller the better; other makes like Sanderson were steadfast in their thinking towards plates. A glance at a Wallace Heaton advert of 1923 shows far more plate cameras for sale than roll film models and yet, you can almost see the niggling doubt in the minds of the plate camera manufacturers. They thought they were right to go on producing their models, but just supposing roll film did win the day... what then?

Film or plates?

It was this thinking that led to the growing popularity of film packs. A flat box containing the pack fitted the back of a plate camera in place of the usual plate holder. The pack contained perhaps a dozen sheets of cut film. Each sheet was attached to

How different camera manufacturers fought a losing battle against roll film. Right, a film pack seen here fitted to the back of an Ica camera and, below, three cameras that offered a choice of plates or roll film. Left to right, they are an Eastman No.4 Series C quarter plate camera with a rather bulky roll film back; an Ensign Regular, essentially a roll film camera but with a plate option; and a St. Etienne Universelle with half-separated film back.

a tab of paper which was numbered and protruded from the top of the pack. As an exposure was taken, the first of the tabs was pulled and the film attached was rolled away from the film plane and round to the back of the pack, leaving the next sheet ready for the following exposure.

Plate cameras of course, were still being made to take roll film backs as well, but strange as it may seem, even the roll film makers didn't seem too sure of themselves.

In their hearts, they still saw the plates gave superior results and so they took out their own form of insurance. Which is why some roll film cameras appeared with a provision for plates.

The Ensign Regular was a perfect example. On the surface, it looked like a perfectly ordinary 120 size roll film camera modelled on similar lines to the Folding Pocket Kodak. But closer examination of the camera's back, would have revealed a small

The Ensign Roll Film Reflex

The Ensign Roll Film Reflex of 1922 was a simple model shaped like a box camera laid horizontally but with a reflex mirror which threw its image onto a ground-glass screen in the camera's top. The screen was shielded by a tall hood twice the height of the actual camera and which folded onto the top when not in use.

One interesting point about this model was the fact that it didn't use the focal plane shutter that was normally necessary in a single lens reflex camera to prevent light from the lens and viewfinder from leaking onto the film. Here's how it worked:-

At the start, the mirror would be flush with the inside of the focusing screen, blocking any light from fogging the film. The movement of a lever on the top of the camera then brought the mirror down to its normal 45 degree angle while, at the same time, opening the shutter behind the lens. The mirror now not only reflected its image up to the ground-glass screen, but also made an effective trap within the body that prevented light from the open shutter from reaching the film.

When the picture had been composed, one of two buttons was pressed on the front of the camera. The mirror flipped up and the shutter snapped shut, all in the fraction of a second needed to make the exposure.

The camera illustrated is a tropical version.

lever that allowed a thin metal flap to be slid aside, revealing a facility for taking $3^{1}/_{2}$ x $2^{1}/_{4}$ in. plates.

But roll film was catching on, gradually taking over more and more plate camera strongholds. The bulky single lens reflex that had been popular since the 1880s and which had held onto its plate design beginnings for years, suddenly began to appear in roll film form. Cameras like the Ensign Roll film Reflex for eight exposures on 120 film, for example.

The Ensign Roll Film Reflex was an unusual front shutter reflex, but it was by no means the first of its type. Newman and Guardia were playing with the same idea back in 1902. Years later, in 1953 Metz were just one manufacturer to use the idea in the Mecaflex, while Pentax waited until 1978 to re-introduce the style, claiming it to be totally new, in a 110 film size single lens reflex.

The Rollei revolution

In 1928, there was another major advance in camera design with the introduction of the first truly compact twin lens reflex. To trace its origins, however, we must go back briefly to 1920 when a firm called Franke and Heidecke was started, initially with a view to the stereo camera market. Their first production was the Heidoscope, a stereo plate camera with reflex viewfinder. It had three lenses, two for taking and one in the middle that projected its image onto a ground-glass screen.

Then, in 1923, the company produced a roll film version of the Heidoscope called the Rolleidescope and, looking at it, you can almost see the reasoning that led to their next design: take one lens away, turn the camera on its end and put the ground-glass screen on the top. The result was the first Rolleiflex.

The first model in 1928 took 117 film, the same size as 120 but giving only six $2^{1}/_{4}$ in. square exposures. The camera had a rim-set Compur shutter speeded from 1 to 1/300 sec and an f/4.5 Tessar lens. Film transport was by the traditional knob and red window method.

The Baby Rollei which first appeared in 1931, took 127 film, giving twelve $1^{1}/_{2}$ in. square pictures on a roll. But because 127 film was only numbered from one to eight, a new method of automatic counting was needed. So the Baby Rollei appeared with crank winding. The usual red window was used to bring the first number into view and thereafter, the correct amount of film was advanced each time the handle was cranked.

Crank winding subsequently became a standard feature of the Rolleiflex and was introduced to the larger $2^{1}/_{4}$ in. square format a year later when a third model - the Rolleiflex Standard - was introduced for 120 film in 1932.

Up until the advent of the Rolleiflex, all reflex cameras, both single and twin lens, had been bulky affairs and twin lens models, by virtue of their extra lens, had always

How one thing inevitably led to another in the Rollei story. Above, the Heidoscope, a stereo plate camera, shows its influence on the first Rolleiflex while, below, the Baby Rollei (right) introduces crank wind to the Rolleiflex Standard.

been the more cumbersome of the two designs. Now suddenly the tables were turned. The twin lens reflex had been transformed from a bulky giant into a compact masterpiece, leaving its single lens cousin looking like an ungainly monster.

Clearly, something had to be done about it, and in 1933, the first truly compact roll film single lens reflex came onto the market. The camera was the Exakta.

Smaller SLRs

Produced by Ihagee in Dresden, the Exakta used 127 film. It sported an f/2.8 lens and a focal plane shutter speeded from 1/25 to 1/1000 sec. The Model B that followed took the shutter speeds down to a full 12 seconds, had a built-in self-timer and was synchronised for flash.

The following year saw the arrival of another new design which is still the rage today - that of the built-in motor drive. Admittedly today's cameras owe their film

The Robot gave the first practical automatic film advance. This is the Robot II.

The Russian-made Sport and the German Kine Exkakta I (Rectangular window).

transport to micro electric motors, whereas the 1930s version relied on clockwork, but the size of a motordrive camera back in those days, was actually often smaller than today's like-minded compacts. Although the first camera with a motor is usually reckoned to be the Debrie Sept, it must be remembered that this model couldn't make up its mind whether it wanted to be a still or a short-run movie camera. The first actually designed as a still camera for 35mm film was the Robot, originally seen in 1934.

The Robot measured only $4^1/_2$ x 3 x 2 in. and was dominated by a large knurled knob on the top plate. This wound a clockwork motor and, thereafter, the film was advanced automatically immediately after the shutter fired. Pictures were 24mm square on standard 35mm film.

In 1935, the Russians came up with the world's first 35mm SLR. The Sport - or Cpnom - was made in Leningrad, offered interchangeable lenses and a waist-level finder. It was, however, extremely rare outside Russia and was considerably overshadowed by another camera from the West.

In 1936, Ihagee produced a 35mm version of their Exakta. It was called the Kine Exakta and it was this camera that is often mistakenly referred to as the first 35mm single lens reflex. Either way, the style had been set for what was to become one of the most popular camera designs ever produced.

8. Clever, compact and complicated

THE 1930s WERE A TIME of tremendous innovation. Between 1930 and 1940, every year seemed to see the emergence of a major name in the history of the camera. It was the age of the miniature, a time when cameras became compact, clever and complicated - and some became too complicated for their own good.

The final breakthrough for colour film came in the 1930s. Although colour photography both on plates and film had been around for a quarter of a century, every other process was finally put in the shade by the arrival of the integral tripack made possible with Kodachrome in 1935.

Destined to become the first commercially successful amateur colour film, this was a reversal material invented by Leopold Mannes and Leopold Godowsky in co-operation with the Kodak research laboratories. Kodachrome first appeared in 16mm size for movie films; 35mm for still cameras arrived in 1936, the year Agfa brought out their own brand of colour film called Agfacolor.

It would be easy to fill the rest of this book with details of the various experiments that led up to the final production of a viable colour film; just as it would be easy to write a lot more about how colour films progressed after 1935. But the aim of this book is to write about cameras rather than processes; so having noted that it was now easy to take colour pictures in everyday cameras, let us return to the subject in hand.

Better quality

It was black and white film, rather than colour film that affected camera design in the 1930s and, with the advent of Kodak's Verichrome and Ilford's Selochrome, films at last began to do justice to lenses. With better quality negatives, camera manufacturers turned their attention to getting smaller and therefore more pictures from every roll of film.

So it was that roll film cameras with their maximum of twelve pictures per roll set out to challenge the economies of 35mm. The result was a new craze for sixteen-on cameras, models designed to take sixteen pictures on a roll of 120 or 127 film by taking the eight-on format and cutting it in half.

The Ensign Cupid - first of the 16-on cameras.

The idea began with the launch of an Ensign film with the letter X marked between each number that enabled the user to stop winding half way between frames. Used in conjunction with a special mask fitted into the film plane, certain cameras could be converted to take sixteen exposures to a roll.

The first camera specifically designed for the sixteen-on format was the strange little Ensign Cupid of 1923 but it was in the 1930s as a direct challenge to 35mm that the style really caught on. By then, special film was replaced by the more practical idea of using standard film in a camera with two red windows in place of the usual one.

Two in one

The idea was that for sixteen-on pictures, the figure 1 was wound into the first window and then, after exposure, it was wound into the second window. For the third picture, figure 2 was wound into the first window and so on right through the roll. Very often, the camera was supplied with a detachable mask for sixteen-on which, when removed, gave the usual eight-on by using only one window.

A perfect example was the Kolibri made by Zeiss Ikon in 1930. At a time when vest pocket cameras were giving way to 35mm, the Kolibri was the perfect compromise. It looked like a vest pocket model but, loaded with 127 film, it gave sixteen exposures, $1\frac{1}{2} \times 1\frac{1}{4}$ in. to a roll. Even so, the Kolibri was fighting a losing battle and lasted only a couple of years.

Kodak set their sights on Boy Scouts and Girl Guides in 1929 and 1931 respectively when they introduced two new versions of their Vest Pocket Kodaks for these two organisations. The company were licensed to incorporate the Boy Scout and Girl Guide emblems into the cameras which were made in blue for the girls and green for the boys. These were the first Kodak cameras sold complete with carrying case and were made to hook onto a Scout or Guide belt.

The following year, in 1932, Kodak acquired the camera works at Stuttgart from Dr. August Nagel and straightaway, his brother Hugo came to England to set up a design and development team at Harrow. It was this team that was responsible for turning out many of Kodak's successful designs of the 1930s including a Six-20 folding camera and, in 1934, the most advanced version to date of the Brownie, a model for 620 film with f/11 doublet lens, two supplementary lenses and three stops.

By 1939, design had been completed for an all-metal range of box cameras but production was interrupted by the Second World War and these Six-20 Brownies were eventually the first cameras to roll off the Kodak production lines straight after the war. Between 1946 and 1959, 9,000,000 were made at Harrow.

Zeiss go into 35mm

But we are getting beyond ourselves. Back in 1932, Zeiss Ikon, previously renowned for their plate and roll film cameras, surrendered at last to the popularity of 35mm. The result was the first Contax which appeared in the same year, and in competition to, the Leica II. The two cameras were similar in many ways. Both took 35mm film, both had coupled rangefinders.

But the Contax was a long way from being a straight copy of its contemporary. It was a very individual camera with individual design ideas, not least of which was its slightly out of the ordinary shutter. It moved vertically instead of horizontally and was made up from a series of metal slats like a roll-top desk. The camera was launched with ten interchangeable bayonet-fit lenses (unlike Leica's screw-fit) and with speeds

The Contax I, the camera with which Zeiss went into the 35mm market and launched as a rival to the Leica II.

Super Ikonta's superior rangefinders

Most rangefinders at this time worked by looking at the subject from two viewpoints, giving two images brought together by a tilting mirror. When the images coincided, the distance was read off a dial. The Super Ikonta, tackled the idea differently.

A tiny glass window was mounted beside the lens. Inside this, were two wedges of glass, layed with the thin end of one against the thick end of the other, so that the result was a parallel-sided piece of flat glass which in no way changed the direction of light passing through it. As the front element of the Super Ikonta's lens was revolved for focusing, so these two wedges also revolved, one in the opposite direction to the other. After 180 degrees of turning, the wedges would end up thick end against thick and thin against thin, giving a prism of glass which changed the light's direction.

It gave a perfect way of swinging one of the sighting lines in a rangefinder and, as the instrument was coupled to the front element of the camera lens, the Super Ikonta ended up with a superior rangefinder.

from 1/25-1/1000 sec. Two years later, a new model was produced with extra slow speeds from 1/2-1/10 sec.

The first Super Ikontas appeared from Zeiss in 1933. They were folding cameras for $3^1/4$ x $2^1/4$ in. pictures that sprung open to the taking position at the touch of a button and which used a unique form of coupled rangefinder. Such an accessory was, of course, not unknown before the Super Ikonta; and rangefinders were already being coupled to lenses in the current miniatures, but the Super Ikonta was the first since the Kodak model to make the idea popular on roll film cameras

Unique designs

There were new models with unique designs appearing on the market all through the 1930s, but there were new variations on old designs as well. The Voigtlander Prominent, also born in 1933 was such a camera. Basically, it was a folder that sprang open at the touch of a button and which took either eight or sixteen pictures on 120 film. But the manufacturers added so many bits and pieces that it took on a weird shape never before seen in the style.

It featured a special light-tight series of openings in the bellows for letting air in and out as the camera was opened or closed, an f/4.5 lens, shutter speeded from 1 - 1/250 sec, coupled rangefinder and extinction type built-in exposure meter. The extra accessories not normally associated with a humble folder added knobs and dials to the bodywork, giving it an ugly appearance that probably accounted more than anything else for its unexpected lack of popularity.

The camera died, but the name lived on, and in the years that followed the Second World War, Voigtlander produced their second version of the Prominent, this time, a 35mm non-reflex model with interchangeable lenses.

But the folding camera wasn't the only design that ran through several different makes. The Rolleiflex was now beginning to show its influence over other manufacturers.

Popularizing 35mm photography - Kodak's Retina I.

In 1933, Voigtlander introduced the Superb, a twin lens reflex with a unique parallax correction device achieved by actually tilting the viewing lens and screen within the camera so that each of the twin lenses always aimed at the exact point of focus. Ironically, the Superb didn't do as well as the Brilliant, a cheaper Voigtlander with the appearance of a twin lens reflex but whose viewing screen was really nothing more than a big condenser lens type viewfinder.

The Ikoflex was another twin lens reflex that came out of the Zeiss works in 1934. The first models had lateral film transport by a lever at the camera's base, but this gave way on later models to vertical transport by the more usual winding knob.

A cheaper Rollei

In 1933 the Rollei works adapted their own design to give us the Rolleiflex's cheap cousin, the Rolleicord. It was a basic twin lens reflex with film transport supplied by turning a knob and operating a counting mechanism. With a simple Triotar lens, the Rolleicord sold for about half the price of a Rolleiflex.

Not that these medium format cameras came anywhere near killing off the interest in miniaturization. The interest was there all right but by now it had become apparent that in some ways, 35mm cameras were defeating their own object. While the film was more economical than standard roll film, the cameras that used it were too expensive for the average amateur. What was needed was a well made, quality 35mm camera that would sell for a fraction of the cost of a Contax or Leica.

As was so often the case, it was Kodak who came up with the answer by launching,

Three miniatures from the 1930s. Left to right, they are, the Ensign Midget, the Eljy and the Coronet Vogue.

in 1934, the first of the Retinas, a folding camera for 35mm with an f/3.5 lens specially designed for the format by Schneider and a Compur shutter speeded to 1/300 sec. It was quality coupled with cheapness that made the Retina series, probably more than any other, responsible for the real popularization of 35mm photography.

But the 35mm format didn't necessarily rely on 35mm film. The mid-1930s saw the birth of a number of interesting miniatures.

Small formats

In 1934, there was the Ensign Midget, an interesting little strut camera measuring no more than $3\frac{1}{2}$ x $1\frac{1}{2}$ x $\frac{1}{2}$ in. when closed. It came at first in two models, one with a fixed focus lens and the other with an f/6.3 Ensar and shutter speeded from 1/25 to 1/100 sec. The camera took six pictures on special Ensign film known as Lukos E10 and the negative size of 30 x 40mm was only slightly larger than the conventional 35mm format.

The 1936 Coronet Vogue was a fixed focus, fixed aperture, single shutter speed model made of Bakelite (used first in the Rajar No.6 of 1929) and designed for use with Vogue V35 film. The camera measured $4\frac{1}{2}$ x 3 x 1 in. when folded and the picture size was 50 x 30mm. And 1937 saw the Eljy from the French firm of Lumière. This was a rigid bodied camera measuring 3 x $1\frac{3}{4}$ x $1\frac{1}{2}$ in. that featured an f/3.5 lens, a shutter speeded from 1/25 to 1/125 sec and which gave negatives 36 x 24mm, the standard 35mm format, on unperforated film.

Smaller still was the Coronet Midget that appeared in 1934 for 16mm roll film. Picture size in this was a mere 18 x 13mm from a camera that measured only $2\frac{1}{4}$ x $1\frac{1}{4}$ x 1 in.

The mid-1930s also saw a strange hybrid of the miniature and the medium format worlds; a 35mm twin lens reflex considered by many to be the pinnacle of perfection both then and for many years to come. It arrived in 1935 and was produced by Zeiss who saw it as being two cameras in one - a reflex and a Contax. Consequently, they called it the Contaflex (not to be confused with the single lens reflex of the same name produced by Zeiss after the Second World War).

The Contaflex was ahead of its time. It was a camera equipped to produce the best ever results on 35mm film. Unfortunately the film at that time wasn't up to handling those results. The format was still considered a bit of a novelty and its grain structure was nowhere near a fine as present standards.

Coronet Midget from 1934.

Stereo photography, while no longer as popular as it had been at the beginning of the Century, still hovered under the surface of camera manufacture, rearing its head now and then with an unusual model of twin lens camera. In 1935, no less a manufacturer than Leitz considered the possibilities of such a camera with a prototype stereo Leica. It was made personally by Oskar Barnack and was designed as two cameras in one with a double shutter. Height and width of the camera were the same as a standard model but the length of the camera was 7¹/2 in., housing two lens flanges 2³/4 in. apart. The size of each stereo frame was 24 x 22.5mm and after each exposure, the film was advanced the equivalent of two frames.

The stereo Leica, however, remained only a prototype and was never actually put into production, even though the company did make several attachments to convert standard models to stereo photography.

Japanese moves

Meanwhile, half way across the world, things were beginning to stir now in the Japanese camera industry. Nippon Kogaku who, for the previous eighteen years, had made only military optical and measuring equipment, turned their attention in 1935 to producing their first camera, of which explicit details have been lost. A second

Japanese company established two years earlier in nothing more than a rented three-room apartment in Tokyo, also produced a camera that year; but the genesis of their model might have started with an event of six years before. One story suggests that in 1929 that the Graf Zeppelin made a stopover in Tokyo. The skipper had a Leica - the first to be seen in Japan and a style that was copied by that small three-room company whose name was The Precision Instruments Optical Research Laboratory.

The Twin Lens Contaflex

The Contaflex, like a few other early twin lens reflexes, featured horizontal film transport across the camera body but, unlike any other, it produced a standard 36 x 24mm negative on 35mm film. While this size was suitable for a negative, however, the manufacturers were swift to realize that it was far too small for accurate focusing. So the Contaflex became the only camera to have a viewing screen with a larger image than the negative area.

This was achieved by using lenses of different focal lengths for taking and viewing. The taking lens was a 50mm in a choice of f/2.8, f/2 or f/1.5. The viewing lens was a longer focal length with the extra light path accommodated by placing the reflex mirror as far back as possible and then moving the viewing screen an appropriate amount higher than usual. The effect was to give the Contaflex a tall, slim look never seen before or since in a twin lens reflex.

There was a range of interchangeable lenses, all of which coupled with the viewing lens when the screen was masked down proportionally and this screen not only had a brilliancy condenser fitted, it also sported a hinged magnifier inside the hood for critical focusing.

As if this wasn't enough, the camera had a built-in exposure meter, equipped with a series of prisms that changed the angle of the sensitive cell's view along with the lens, a metal focal plane shutter speeded from 1/2 to 1/1000 sec and a delayed action feature.

Two very different firsts from the 1930s. Left, the Reflex Korelle, first 6 x 6 cm SLR; above, the tiny 16mm Minox.

Whether the Zeppelin story is true or not, the camera was an obvious Leica copy. It was named the Kwanon, after the Buddhist Goddess of Mercy, but built only as a prototype. By the time the company had produced a model for general sale, religious connections had proved unacceptable and the name was changed to Canon.

The following year, 1936, saw the arrival of the first $2^1/4$ in. square single lens reflex. The German Reflex-Korelle sported an f/3.5 Schneider-Kreuznach lens and focal plane shutter speeded from 1/10 to 1/1000 sec. The camera was sold again after the Second World War as the British Agiflex.

The same year also saw the introduction of an unusual little camera called the Purma. With an f/6.3 fixed focus, fixed aperture lens, the camera had three shutter speeds controlled, strangely enough, by the way it was held. Horizontal, it gave 1/50 sec, but if the photographer turned the camera on its side he would hear a quiet click that indicated that the shutter would now operate at 1/25 sec. Held on its opposite side, the speed increased to 1/450 sec. The device was known as a gravity shutter and the pictures produced naturally had to be square - sixteen on 127 film.

Complicated masterpiece

At the beginning of this chapter we mentioned that some of the cameras that appeared in the 1930s were too complicated for their own good. A perfect example came in 1937 from Swiss watch and clock manufacturer Le Coultre and designed by Englishman Noel Pemberton Billing. The camera was the Compass, an intricately beautiful masterpiece of miniature engineering that seemed doomed from the start.

A miniature that did survive, however, was the Minox, made by Valsts Elektro-Techniska Fabrika in Latvia and seen for the first time in 1937. The Minox rapidly

The complicated Compass

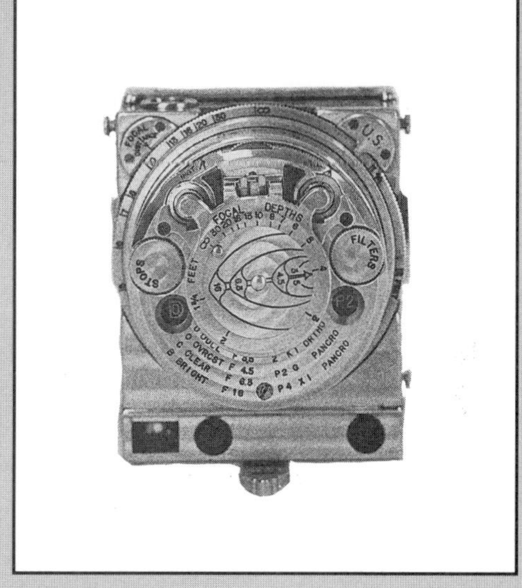

The revolutionary little Compass featured an f/3.5 lens with rotary shutter in twenty-two speeds from 1/500 to four-and-a-half full seconds. It had a rangefinder in a separate window to the viewfinder which, in turn, had a pivoting mirror for right-angle viewing.

The body was metal, incorporating a device for taking stereo or panoramic pictures. Built-in yellow, green and orange filters were dialed in from the side of the lens. Also featured were a ground-glass screen with focusing magnifier, collapsible lens hood, hinged lens cap and built-in spirit level, built-in exposure meter and depth of field scale. It took single pictures on plates or cut film and also had a roll film back.

Now you know its specification, take a look at its size: folded, the camera measured no more than $2^{3}/_{4}$ x $2^{3}/_{4}$ x $1^{1}/_{4}$ in.

The Compass was probably one of the most beautiful little cameras that has ever been produced. Unfortunately, the very nature of its beauty was its downfall. Put its specification on a camera five times the size and you have an extremely worthy piece of equipment. Put the same specification on a miniature and all those accessories and controls are just too fiddly.

Their size made the photographer feel like a clumsy oaf and the camera was too expensive. With the outbreak of the Second World War, the Compass died away, never to be seen again.

became epitomized as the ideal spy camera, a role which it continues to hold today if television and films are anything to go by. It was a perfect example of the really practical miniature, taking fifty 8 x 11mm pictures on 9.5mm unperforated film, loaded into special cassettes, all housed in a slender body measuring 3^1/4 x ^{3}/4 x ^{1}/2 in. The lens unusually, had a fixed aperture of f/3.5 with the exposure controlled by shutter speeds ranging from 1/2 to 1/1000 sec. The shutter was cocked and the film advanced each time the camera was slid closed and opened again.

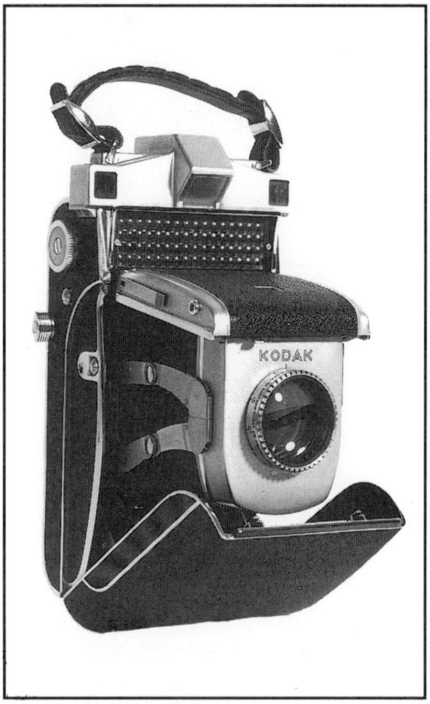

The Super Kodak Six-20 - first with automatic exposure control.

Auto exposure

The first camera to feature automatic exposure control appeared in 1938 and with an already impressive track record, it isn't difficult to guess the maker. Once again, Kodak was first in the field.

The camera was the Super Kodak Six-20, a folding bellows camera for eight exposures on 620 film. It had a shutter speeded from 1 to 1/200 sec, coupled rangefinder and apertures from f/3.5 to f/22 that were varied automatically according to the light being read by a photo cell spread across the camera's width above the lens.

Kodak's brief had always been to make photography simple while bringing advanced techniques within the range of the man in the street. With the Super Kodak Six-20, thcy again proved their point. For the first time, Mr. Average could buy a camera that was technically competent of producing worthwhile results under varying conditions but which, unlike its predecessors, assured the photographic layman of a correctly exposed negative every time.

9. Instant pictures

THE STORY GOES that it started with a father taking pictures of his young daughter; it ended with a giant step in the history of the camera. The father was Dr. Edwin Land and his daughter, having had her photograph taken, became upset when she learned how long she would have to wait to see the results. The little girl's impatience led her father to begin wondering about the possibility of building a camera that not only took a picture, but also developed it instantly within a minute of the exposure being made. The result was the first Polaroid camera and the rest is history.

Land was a fanatic on the theory and application of polarising filters, an interest that he seems to have developed in 1926, almost as soon as he arrived at Havard at the age of 17. He in fact left the university the same year and started pursuing these interests independently, with particular emphasis on using polarising filters in motoring to reduce headlamp glare in opposing lines of traffic. Two years later, at the age of 19, he made a patent application for a polarising material for that purpose.

Edwin Land, inventor of the first commercially successful instant picture camera.

Polaroid Corporation

His real innovation was to come up with a method of making polarising filters in sheet form and taking its name from his invention, the Polaroid Corporation came into being in 1936. By 1945, when the possibly apocryphal story of the father and daughter exchange took place, he was an acknowledged expert on the subject of polarised light and a man of rare innovative thinking. He therefore had the right kind of mind to begin thinking seriously about making his daughter's dream a reality. Within six months he had made an application for a patent and, in 1947, the first demonstration of instant photography took place at a meeting of the Optical Society of America.

The first company - reputedly Kodak - that Land approached to turn his system into a commercial proposition turned the idea down flat. It was nothing more than a gimmick, they said, and one without a future. In 1948, the first instant picture camera appeared. Fourteen years later, 14,000,000 had been sold and the Polaroid Land Corporation was the second largest producer of photographic materials in America.

The Polaroid process

Instant pictures were achieved by use of Polaroid Land film which consisted of a negative emulsion and positive paper on two separate rolls, together with pods of chemicals. As each picture was taken, the photographer pulled a tab which led the exposed negative, the positive paper and a chemical pod through two tight rollers. The pod broke and released the chemical which was spread between the negative and positive sandwich.

It was this chemical, rather than light as would have been the normal practice, that transferred the image from one to the other. The positive image was formed on the paper with the help of unexposed silver crystals from the negative.

The first Polaroid

Edwin Land's first instant picture camera was known as the Model 95. It was a folding design but somewhat larger than the majority of roll film cameras of that time. It had an f/11 lens and exposure was simplified by coupling the aperture to the shutter speeds, giving combinations from 1/8 sec at f/11 to 1/60 sec at f/45.

The pictures were sepia, but in 1950, a new black and white orthochromatic film called Type 41 was introduced. Two years later, the first instant camera for professional use appeared on the market. It was the Model 110 with an f/4.5 lens and shutter speeds of up to 1/500 sec.

Polaroid went on extending their range right throughout the 1950s, and in 1958, they took their process a step further by producing separate Polaroid backs to fit conventional cameras. The back consisted of a holder that took film packets with the negative-positive sandwich and processing pods enclosed within a light-tight envelope. For processing, the envelope was pulled through the usual rollers and all the way out of the holder, where it continued developing inside the black envelope.

The Polaroid story is one that has never stood still and, as we shall see later, they went on to develop more and more sophisticated forms of instant photography which in 1963 inevitably led to instant colour. Of which, more later.

Eye-level SLR

The late 1940s and early 1950s have, however, gone down in the history of the camera for more than the birth of instant photography. The first eye-level, single lens reflex was exhibited in 1948. With the Contax S - the first miniature with a pentaprism - the photographer could hold the camera to his eye and see exactly what he was photographing, the right way round and the right way up. The camera also differed from previous Contax models by replacing the old metal focal plane shutter with one of cloth and by changing the lens mount from bayonet to screw fitting. The Swedish Hasselblad 1600F was introduced the same year. Two cameras, both single lens reflexes, one for 35mm, the other for 6 x 6cm; between them, they revolutionised their respective formats.

Stereo cameras enjoyed a brief revival in the 1950s with models like the Iloca Rapid

Two single lens reflex firsts. Left, the 6 x 6 cm Hasselblad 1600F; above, the 35mm Contax D.

The Wrayflex and Periflex

The end of the 1940s and early 1950s saw not only innovative designs, but often some downright eccentric ones.

Take the Wrayflex (right), for instance. It was a 35mm single lens reflex, but one which featured the viewing system on the *bottom* of the body, so that the camera was held pressed against the forehead - in theory, possibly better than squeezing it against the nose, but a style that never really took off. The Wrayflex didn't use the usual SLR five-sided prism for viewing the image in the viewfinder either. Instead, it used mirrors.

This meant that when the camera was held horizontally, the image was the right way up, but reversed left to right. With the camera held vertically, the viewing image was the right way round, but upside down!

Another unusual viewing system was to be found on the Periflex series of cameras which began in 1953. The original Periflex (below left) was a British camera conceived by Kenneth Corfield, which used the same screw and register as Leica, so it could be used as a cheap second body for Leica owners.

The Periflex's unusual feature was a kind of miniature inverted periscope which was lowered into the film plane before shooting. This picked up just a small area of the actual image for focusing. The device was removed before the exposure was made and an optical viewfinder used for composition. The method gave extremely accurate focusing, but of such a small part of the image, that it was difficult to see what was actually being viewed!

that had f/2.8 lenses, Stereo Pronto SVS shutter speeded from 1 to 1/300 sec and a coupled rangefinder. Slightly more unusual from the early 1950s, was the Italian ISO Duplex which used 120 film widthwise, giving two images 24 x 24mm square across the film. The lenses were f/6.3 and shutters were speeded from 1/25 to 1/100 sec.

But, as had so often been the case before, the stereo craze was brief and soon died,

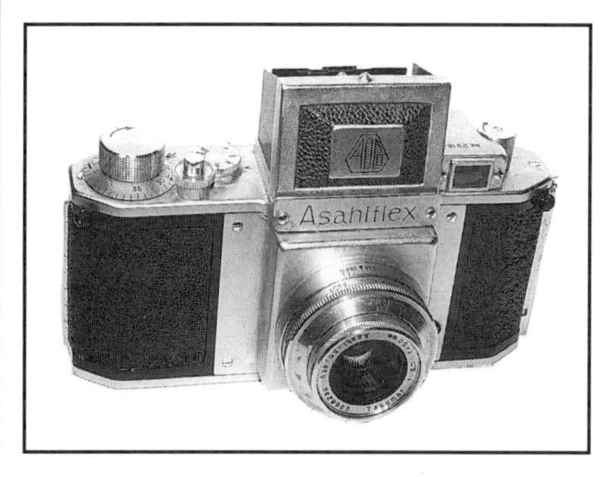

Two firsts from Japan. Left, the Asahiflex and, right, the Nikon I.

even though the Russians brought out a plastic model as late as 1960. It was called the Sputnik and had not two lenses, but three; the two outside lenses for the stereo pair and the one in the middle and slightly above the other two for a ground-glass screen. All three were geared together for focusing.

In 1952, after seven years of research, the first 35mm single lens reflex came out of Japan. But despite the influence of the eye-level Contax, it stuck to the rather more tried and tested formula of the waist-level viewfinder. The camera was the Asahiflex, made by the Asahi Optical Company. It had a 50mm f/3.5 Takumar lens and a focal plane shutter speeded from 1/25 to 1/500 sec. As well as its waist-level viewfinder, it also featured an eye-level optical viewfinder, though this of course was not reflex.

Two years later, Asahi claimed their own first. Until then, it was usually accepted that single lens reflexes suffered from the picture in the viewfinder blacking out after the exposure had been made, as the mirror flipped out of the light path and failed to return until it was physically lowered into position. In 1954, the Asahiflex II appeared with what is often thought to be the first instant return mirror. Granted that the camera I was probably responsible for popularizing the style, but the instant return mirror *first* appeared on the Vanneck of 1890 and the style was also seen on 1947 Duflex.

In 1957 Asahi designed their first camera with a built-in pentaprism and named it after a corruption of *pentaprism* and *Asahiflex*. The result was the first Pentax.

The first Nikon

The span of years between 1948 and 1957 was also important for another Japanese camera manufacturer. Nippon Kogaku released the Nikon I in 1948. In appearance and specification, it was similar to the pre-war Zeiss Contax, but with a film format of 32 x 24mm, rather than the more usual 36 x 24mm, a fact that the then current open house on patents made perfectly permissible. In 1955 the first 35mm Minolta arrived in Britain from Japan. Pre-war there had been Minolta copies of Zeiss Ikontas, but the 1955 camera was a 35mm rangefinder model with a fixed lens called the model A.

But it was probably the Korean War more than any other factor that made Japanese cameras in general and Nikons in particular known to the western world. By 1952, the Far East was crawling with western newsmen covering the war and, finding no repair facilities for their Leicas and Contaxes in the East, they turned to the locally produced goods. It wasn't long before they realized they were on to a good thing.

Choice of lenses

By the late 1950s, the Nikon S1 to S4 rangefinder models were in wide use, offering a choice of over thirty lenses from 25mm to 1000mm and with a 50mm standard lens that featured the previously unheard of wide open aperture of f/1. There was also available an equally revolutionary motor drive that gave the user the benefit of four frames per second shooting.

Perhaps the most famous of the range was the Nikon SP from 1957. It featured a

The Mecaflex

In 1953, what is now a very rare camera made its debut. The Mecaflex was based on a 1947 prototype from Heinz Kilfitt of Munich and made in the 1950s by the Metz radio and TV maker. The Mecaflex was an extremely compact, albeit heavy, 35mm single lens reflex, measuring no more than 4 x $2^1/2$ x $2^3/4$ in.

Closed, it boasted extremely smooth lines, but then the whole of the top plate hinged up to reveal a small waist-level viewfinder, with magnifier, a lever film wind, the shutter release and the film rewind knob which also doubled as a film reminder.

Lenses were interchangeable, but

the only ones that seem to be known are the 40mm f/3.5 or f/2.8 standards. The behind-the-lens (not focal plane as in most SLRs) shutter was speeded from 1-1/300 sec. and the film format was 24 x 24mm, offering 50 exposures on a standard cassette of 35mm film.

The Nikon F, a firm favourite with professionals right from the start. It was the first 35mm SLR designed to take a motor drive.

combined viewfinder and rangefinder window with fields of view for different focal lengths of lens from 28mm to 135mm plus parallax correction for normal and tele lenses. A battery-operated motor drive was a featured accessory, offering automatic film advance at three frames per second. The shutter was speeded from 1-1/1000 sec and the standard lens was either an f/1.4 or f/2 50mm Nikkor.

Nikon was probably one of the first companies to recognise the importance of being able to shoot at several frames per second with a camera, so much so that when they introduced their first single lens reflex in 1959, it was designed right from the start to accept motor drive. It was this model, more than any other, that made Nikon a household name with professional photographers worldwide. The camera was the Nikon F. It lacked a metering system of course in those days, but a Photomic viewfinder, available in 1962, could be attached which sported its own photo electric cell linked to the shutter and diaphragm. Over a million Nikon Fs were made, the camera remaining in production until 1974.

Box camera innovations

While the precision miniatures were growing more professional every day, there was, as always, a strong market for the casual snapshotters, and for them box cameras were still the order of the day. Even in the mid-1950s, Kodak were still launching new models. Their Six-20 Brownie Model F was about as sophisticated as any simple box camera could ever hope to be.

The camera was finished in a buff leather cloth with a brown metal front plate. Large, clear viewfinders were surrounded by frames of polished brass, while both the shutter release and film wind knob were of white plastic. Aperture and focus were fixed but the shutter could be moved from instantaneous to brief time at the flick of a lever. There was a supplementary lens built in for subjects closer than ten feet and a built-in yellow filter as well as facilities for cable release and flashgun.

But by then, it must have been clear that the box camera was finally on the way out as simple cameras began taking on shapes never before seen. The Ensign Ful-Vue

The evolution of the Ensign Ful-Vue, from box camera, through the standard model to the Super. The model was also available in a variety of colours as well as black.

Super for instance. If current advertisements are to be believed, the original Ful-Vue 'revolutionised photography in 1939' and with the Super version released in 1955, the advertisers were ready to claim: *'The Ful-Vue Super looks and performs like an expensive twin lens reflex.'*

Well true it did look a little like a twin lens reflex, having two lenses mounted on the front. But the top lens was there only to provide an image for the extra large viewfinder. Its unusual shape was quite streamlined but its specification of a single speed shutter and fixed aperture made it really no more than a box camera with a super viewfinder. True the lens could be focused from two metres to infinity, but the focusing movement was in no way coupled to the viewing lens and it would have taken a good stretch of the imagination to compare the camera with a real twin lens reflex such as the Rolleiflex.

Flash synchronisation

Even cameras that were called box cameras were beginning to stray away from the traditional shape. The Ilford Envoy Box Camera was synchronised for flash and had a direct vision viewfinder.

A similar style was seen in the Agfa Clack that sported two apertures, a built-in portrait lens and a choice of instantaneous or brief time exposures, as well as flash synchronisation; or in the equally similar Kodak Six-20 Brownie Special.

The 1950s also saw the change from screw to bayonet mount lenses for Leica. The

Thc first of a new generation. The Leica M3 switched Leica from the old screw-fit lenses to a bayonet fit.

new style was launched with the M3 in 1954. Other changes in the new-look Leica included a lever wind for film, rather than a knob (still seen on the classic Leica IIIg up to the end of the decade), all shutter speeds were incorporated in a single dial and the parallax-corrected brightline viewfinder sported frames for 50mm, 90mm and 135mm lenses, each frame automatically selected as the appropriate lens was fitted. A simpler version, the M2, was launched in 1957 without a self-timer, but with an extra viewfinder frame for 35mm lenses; the Leica M1 of 1959 was simpler still with no rangefinder or viewfinder frame for telephoto lenses.

It was around this time that Kodak, as ever with their eye on the snapshooting man in the street, introduced a camera that more than any other hammered home the last nail in the box camera's coffin. The camera was the Kodak Brownie 127.

The extremely popular Kodak Brownie 127.

Smart and inexpensive, with an eye-level optical viewfinder, single speed shutter and fixed focus, fixed aperture, meniscus lens, it was one of the least sophisticated cameras around. Yet its ease of operation was surpassed only by its popularity. Suddenly, box cameras that had been in the family for years were relegated to the loft as snapshotters everywhere began buying the Brownie 127.

The 1950s were a time of change, but they were also a time when many old styles lingered on. At the beginning of the decade, plates and film packs were rife; by the end, the 35mm single lens reflex had established itself as the professional tool. And yet, in 1955, there were still manufacturers turning out brass and mahogany plate cameras with separate shutters which could easily have passed for models from the turn of the century.

First with a built-in electronic flash - the Voigtlander Vitrona. The long handle incorporates the batteries needed to power the flash.

Old designs

The Aptus is a case in point. Looking at such a camera now, who would believe that it came from the same decade as the Nikon F? Made by Moore and Company of Liverpool, the Aptus had been around from as far back as 1895, when it was used for ferrotype photography. But the latest version that was being made and sold in 1955 operated on a new system that worked like this:-

Anything up to 100 sheets of a special sensitive material on an aluminium foil base were loaded into a magazine in the base of the camera. These were taken out one at a time and held in the film plane by a weird and wonderful suction-operated holder. Once the exposure had been made, the sheet was dropped straight down into a special developing tank clamped on the base of the camera. With development completed, the tank was swivelled through 180 degrees and the developed sheet removed to be fixed in a separate tank.

The operator now had a negative which he slotted into a copying arm that incorporated a supplementary lens and the whole thing was mounted on the front of the camera. The negative was copied and thus the new negative, when developed in the same way, became the finished positive.

In 1957, Kodak took their simple cameras another step forward and introduced the Brownie Star series, the first low-priced cameras designed specifically to take colour slides as well as colour and monochrome prints. One of the series, the Kodak Brownie Starflash, became the first Kodak camera to feature a built-in flashgun.

The history of the camera is liberally sprinkled with early examples of ideas that would later become claimed as original. We've already seen many of these, and 1957 saw another with the launch of the Minolta 16, a forerunner of the 110 format that wasn't due to arrive on the scene for another 14 years.

What made the camera so much like Kodak cameras of later years was 16mm film, used in double cassettes, ready to drop into the camera without the need for complicated loading procedures. Transport of the film between exposures was effected by closing and opening the camera, à la Minox.

The simplest of the Minolta 16 range was the Model P with fixed shutter speed and f/3.5 lens. Later models had f/2.8 lenses, a range of shutter speeds and even an exposure meter, coupled directly to the lens apertures.

Another forerunner of one of today's most popular styles of camera was announced in 1959. It was the Voigtlander Vitrona, the first 35mm camera with a built-in electronic flash. Looking basically like any other non-reflex 35mm camera of the day, it had the addition of a huge, ugly handle beneath the baseplate, nearly twice as long as the camera's actual depth. This housed the small but extremely powerful batteries needed to fire the flash. It wasn't a camera to take out in the rain!

By 1962, Kodak were beginning to think about a whole new type of camera, one that would once again prove their ability to be first in the field with revolutionary new styles. A year later, the first Kodak Instamatic appeared and a new era in cameras was born.

10. The electronic revolution

THREE MAJOR INNOVATIONS epitomize the 1960s: electronics, automation and mass production.

Between 1959 and 1961, the history of the camera took two giant steps forward at the Canon plant in Japan with the introduction of the Model P and Canonet. Up until the Model P, camera manufacture was a job for experts and good quality equipment was made by hand, one camera at a time. But at Canon, camera making was getting a new look. Production was streamlined, parts standardized and for the first time a conveyor belt appeared in the workrooms. The Model P was the world's first high-quality camera to be mass produced on an assembly line.

Price was the breakthrough which made the Canonet so remarkable. It was launched at the 1960 Photokina as a good quality camera which could be bought with one month's salary - a good deal less than comparable cameras of the day. The Canonet was a 35mm model with coupled rangefinder, but the thing that set it apart and gave it a unique look was the design of the meter cell which circled the lens, giving an automatic correction for filters which fitted over the whole assembly.

This first entry by Canon into the popular market received so much publicity that

Two important Canons from this era. Left, the Canon P, the first quality camera to be mass produced; right, the Canon 7 with f/0.95 'dream lens'.

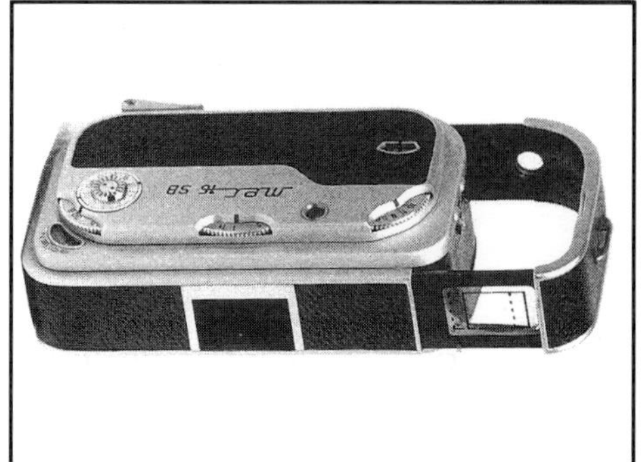

Two firsts for through the lens metering. Left, the Topcon RE Super, first SLR with the feature; right, the Mec 16 SB, the first camera to actually offer the feature.

buyers are reputed to have literally fought at sales counters to buy the camera when it went on sale a year later.

In the same year at Polaroid, Dr. Edwin Land gave a progress demonstration of instant colour at the annual shareholders' meeting, taking three pictures in one of the company's regular cameras. It was another three years before his instant colour film went on worldwide sale. Also in 1960, Polaroid announced the Model 900 Electric Eye Camera, its first fully automatic model and development time of instant black and white pictures was cut from one minute right back to between ten and fifteen seconds.

A year later Canon announced a remarkable lens made for the company for their new Canon 7. The focal length of what was known as the 'dream lens' was a standard 50mm, but the aperture was an astounding f/0.95 - more sensitive even than the human eye, if advertisements of the time are to be believed.

Through the lens metering

Japan was the next country with a significant landmark in the history of the camera. It came in 1963 in the shape of the Topcon RE Super, the camera usually credited with being the first model with through-the-lens metering. What isn't generally known is that the Topcon RE Super was predated by at least two years by another through-the-lens metering camera, a 16mm sub-miniature made by Feinwerketechnik of Germany. The Mec 16 SB sold in England in 1960 and utilized a Selenium cell behind the lens; a fact that the instruction book was swift to point out meant no more worries over filter factors as the meter took its readings through both lens and any filter which might be attached.

The Mec, of course, was a non-reflex camera and, in all fairness to the Topcon company, it was their RE Super, a top quality 35mm single lens reflex, that brought through-the-lens metering to the attention of the world and set a style that was rapidly followed by just about every other major manufacturer.

By now, of course, the 35mm single lens reflex had been accepted by professionals and serious amateurs as the ideal camera for compactness, convenience and versatility. But that still left one very large section of the camera-buying public who would never have the need for such sophistication. The time was again ripe for something revolutionary in the snapshooting market and, as usual, it was Kodak who supplied it in 1963 with the first Kodak Instamatic.

The first Instamatic

In Kodak's first Instamatic, film loading was greatly simplified by the introduction of a pre-threaded take-up spool built into a plastic cartridge. The camera could then be loaded by simply opening the back, dropping in the cartridge and snapping the back shut again. The first Kodak Instamatic for the 126 size cartridge had a single speed shutter and three-element plastic lens. It was synchronised for flashcubes three years later.

Instant colour

But 1963 was famous for more than this Kodak first. That year also saw the arrival of the long-awaited Polacolor Land film for instant colour pictures in one minute flat. Like its black and white counterpart, this was a peel-apart process and worked like this:-

The negative material was made up of three subtractive black and white parts, each coated one on top of the other. In each of these parts, there was a differently sensitive silver halide emulsion which reacted to one of the primary colours, and this in turn was linked to a dye developer of the complementary colour (blue-sensitive layer over yellow dye developer, green over magenta, red over cyan). This pack also included barriers to keep the appropriate developer mixing with its own layer and none of the others.

The receiving layer of the Polacolor film was a water-swellable polymer that contained a substance that fixed the dyes and so produced a positive image. There was a built-in self-washing stabilizing system and, after separation from the negative, the finished positive quickly dried with a hard glossy surface.

Electronic shutters

The idea for an electronic shutter is a lot older than Polaroid and Yashica, each of whom is usually associated with its beginnings. There was a German patent for an electronic shutter as far back in 1902 which relied on a diaphragm driven by an electric motor controlled by a variable resistor.

In 1934, there was an idea going around for another type that relied on a solenoid to open the blades and incorporated a spring to close them again after the solenoid had been turned off.

In 1955 and 1956, both Compur and Prontor took out patents for electronic shutters operated and controlled by photoconductive cells through locks and electronic trigger circuits.

What is not always appreciated when singing the praises of Polaroid is the fact that the negative section of the early peel-apart Polacolor was made by Kodak and a few years later, in 1969, they were all set to launch their own instant photography cameras. Then came a whisper about a completely new instant colour system from Polaroid known as SX-70 and Kodak scrapped their plans. In the end, they launched their own version of SX-70 in 1977. Of which, more later.

Along with Polacolor in 1963, Polaroid also launched their first colourpack camera, but its ability to produce instant colour pictures wasn't its only interesting feature.

Early electronics

The Polaroid Automatic 100 was the first camera ever made with a transistorized electronic shutter for completely automatic exposure control under all lighting conditions, even though Yashica can lay claim to the first true electronic shutter - admittedly not linked to an auto exposure system - in their TL Electro-X of 1958. The shutter for the Polaroid camera was designed, not surprisingly then, by Yashica.

Still with Polaroid, the company introduced the Swinger, their first low-priced instant camera, two years later in 1965. But while the advantages and wonders of instant photography should not be decried, it must be admitted that for the serious professional, everything else that year was overshadowed by the arrival of one other long-awaited camera that had been anticipated for years. It was the year of the Leicaflex, arriving in the shops a year after production had begun in 1964.

There was another slightly unusual model launched by Canon in 1965. This was the Pellix which, to all outward appearances appeared to be a perfectly normal single lens reflex. It was on the inside that things were different. Whereas the conventional

The first Leicaflex

Leica's first single lens reflex was introduced together with four new lenses of the most popular focal lengths and all featuring automatic diaphragms. The Leicaflex sported a newly developed focal plane shutter capable of giving 1/2000 sec and electronic flash synchronisation at 1/100 sec. The viewfinder image was extra bright compared to many of its contemporaries and featured a microprism focusing aid in the centre.

single lens reflex employed a mirror which flipped out of the way at the time of exposure, Canon fitted the Pellix with a stationary pellicle mirror that split the light from the lens between the film and the viewfinder, eliminating mirror shock and making the Pellix the first ever single lens reflex without the usual viewfinder blackout at the moment of exposure. Its only drawback was that with the light split between viewfinder and film, the viewfinder itself was about one-third darker than a comparable camera and approximately one-third of an f-stop less light fell on the film.

The panoramic camera idea was revived again at the 1966 Photokina in the shape of the Russian made Horizont. It took a standard cassette of 35mm film, on which it gave twenty 24 x 58mm frames, each showing a 120 degree angle of view. The lens, which moved horizontally across the camera during exposure somewhat like the old Kodak Panoram of 1899, was an f/2.8. The shutter was speeded at 1/30, 1/60 and 1/250 sec.

By the latter part of the 1960s, automation was becoming more than just a method of helping the photographer to find the correct exposure; more and more, automation itself was actually setting that exposure, freeing the photographer from technicalities and leaving him more time to concentrate on the artistic side of his craft.

Canon Pellix - SLR without the usual moving mirror.

Panoramic cameras have always intrigued photographers and the 1960s version was the Russian-made Horizont (right) that produced 24 x 58mm negatives or colour slides (above) on standard 35mm film. The Japanese Widelux, made in 1959, followed a similar style. The same idea had also been used much earlier in the Kodak Panoram.

The first automatic camera, as has already been noted, was the Super Kodak Six-20 of 1938, but again there had been a patent as far back as 1902 in which a light-measuring system in a camera automatically adjusted the shutter. The Voigtlander Vitomatic, made in 1958, was a fully-automatic camera, but it wasn't until 1968 that the first fully-automatic single lens reflex appeared on the market and, as has been seen before, it was this design that usually set the style for cameras to come.

Automatic SLRs

The first automatic single lens reflex, then, was the Konica Autoreflex, a shutter priority model with through-the-lens metering. The photographer set the shutter speed between 1 and 1/1000 sec and the camera automatically chose and set the correct aperture. The camera was supplied with a 52mm f/1.8 lens.

The Konica, of course, still used a mechanical shutter. The Yashica TL Electro-X, produced in 1958, was the first single lens reflex with an electronic shutter but it wasn't until an electronic shutter was linked with automatic exposure in a single lens reflex, that the full potential became obvious. The first camera to combine all three was the Asahi Pentax ES which appeared in 1971. As the first automatic single lens reflex with an electronic shutter, it set the style for a whole new breed of camera.

The Asahi Pentax ES was an aperture priority model which meant, of course, that the photographer set the aperture and the camera chose and set its own shutter speed. The shutter was in fact half electronic and half mechanical, the first blind opening mechanically and the second blind being closed electronically after the correct exposure had been given. It was speeded between 1/1000 and a full eight seconds with mechanical speeds between 1/1000 and 1/60 sec. The camera was supplied with a choice of f/1.4, 50mm or f/1.8, 55mm Takumar lenses.

Two landmarks in SLRs of the early 1970s. The Pentax ES (above) and the Canon F1.

The same year saw the introduction of the Canon F-1; but this was more than just a camera, it was a whole system. Until then, accessories tended to be something which grew almost willy-nilly; the F-1 system was designed as an entity - camera, accessories, everything for the complete professional outfit. Right from the start, the F-1 was planned as a part of a complete system, and Canon saw four main accessories as indispensible for their new camera: motor drive, automatic exposure

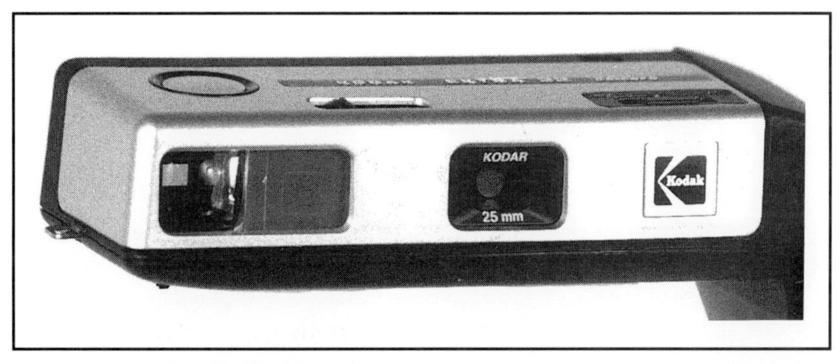

The 110 format came from Kodak in 1972 and remained popular for a decade. This is the Kodak Ektar 22.

The Olympus OM-1. It was first of the compact 35mm single lens reflexes and destined to set a style followed by all the other major manufacturers over the following few years.

control, a booster for the meter and a 250-exposure film chamber.

In 1972, Kodak took their Instamatic design and made it even more portable by reducing the size of the cartridge from 126 to 110. The result was the ultra small Pocket Instamatic cameras which were produced in seven models. Within a very short space of time, most of the other major manufacturers were hot on Kodak's heels, producing 110 cameras of their own. None of the other companies' products could be called Instamatics though. Instamatic was a Kodak trademark and only their cameras could be referred to by that name.

Compact SLRs

The early 1970s were an interesting time with quite a few surprises in variations on old themes. One such revolutionary idea came from Olympus, who took the concept of the standard 35mm single lens reflex and made it about one third smaller than any other make. The result was the Olympus M-1, released at the 1972 Photokina. At that time, the camera was the smallest, lightest, full-frame 35mm single lens reflex in the world, launched as part of a system that featured 280 accessories including thirty-eight interchangeable lenses. Unfortunately, Leitz objected to the name, so Olympus had to relaunch the camera as the OM-1, on sale first in 1974

The OM-1 took the competition by storm. Some laughed and said the size would never catch on; photographers, they said, like something heavier and more 'meaty' in their hands. They didn't laugh for long. Within the next few years, just about every top single lens reflex manufacturer was producing his own compact SLR camera.

By now, electronics were becoming an accepted part of camera manufacture, but in 1972, even they took a new leap forward. The new revolution was a tiny indicator light known as a light emitting diode or LED for short. This device, which is practically everlasting, is in effect a simple diode made of a special semiconductor material such as gallium-arsenide-phosphide which, when attached to a low voltage of around 1.5 volts, lights up with a red or green light.

These LEDs were so small that it became feasible to actually insert them into the camera viewfinder where they could be used to indicate the danger of over or under-exposure, or used in a bank to show the aperture or shutter speed chosen by an automatic system.

The first camera to use LEDs in this way appeared in 1972. It was the Fujica ST 801, again soon to be followed by other leading manufacturers within the next few years.

One-step photography

'Absolute one-step photography', as Dr. Land called it arrived in 1973 with a completely new instant photography system from Polaroid. It was known simply as SX-70 and, unlike its peel-apart predecessor, it involved no waste paper. With one of the special film packs loaded into the Polaroid SX-70 Land Camera, instant colour

Polaroid SX-70 instant SLR

The Polaroid SX-70 Land Camera was a single lens reflex unlike any design ever seen before which used ten-exposure film packs.

Inside the pack, apart from the film, there was a wafer-thin battery to power all the camera's functions, including a motor that drove the reflex mirror and ejected the prints as well as providing power for a flashgun when fitted.

With the SX-70, therefore, the photographer didn't have to worry about fitting new batteries. Folded, the camera measured no more than 7 x 4 x 1 in.

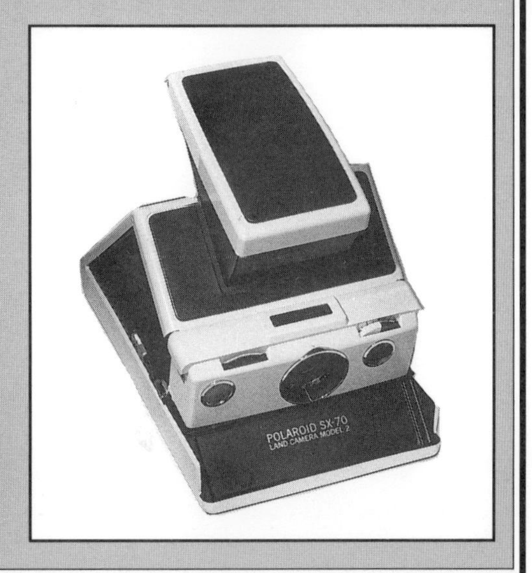

pictures were ejected from a slot in the camera's base less than two seconds after the shutter release was pressed. These prints then developed automatically in daylight before your eyes, drying into a hard, flat glossy print 8cm square.

Computers, very much a growing thing of the 1970s, finally had their effect on cameras in 1976 with the introduction of the Canon AE-1. The versatility of a computer is based mainly on its Central Processing Unit (CPU) and the AE-1, a shutter priority camera, was the first to have one such unit built in. The result was a sophisticated electronic system that enabled the camera to sense and transmit all information to the CPU which in turn controlled the automatic exposure system.

The CPU even controlled flash photography. With the Canon's special Speedlight 155A fitted, the shutter speed was automatically set to 1/60 sec, while the CPU assessed the light from the flashgun in any conditions and automatically set the camera's aperture to the right exposure - all in the brief space of time that it took for the flash to fire.

Canon AE-1 with its own Speedlight flashgun and motor drive, first SLR to use the computer chip to the full.

Instant Kodak

In 1976, Kodak at last launched their PR-10 instant photography system with the introduction of their EK4 and EK6 instant cameras, working on a similar system to Polaroid's SX-70. The launch didn't best please Polaroid, who began court proceedings, destined to go on for the next twelve years, alleging infringement of patents.

The biggest difference technically between SX-70 and PR-10 film was in the fact that Polaroid used emulsions producing their latent image when the film was exposed to light through the front, while in the Kodak system, light from the lens hit the back of the film and the final result was viewed from the opposite side.

The Throwaway Camera - destined to become a craze in the late 1980s - arrived the same year. Made by Rank and used essentially for the premium

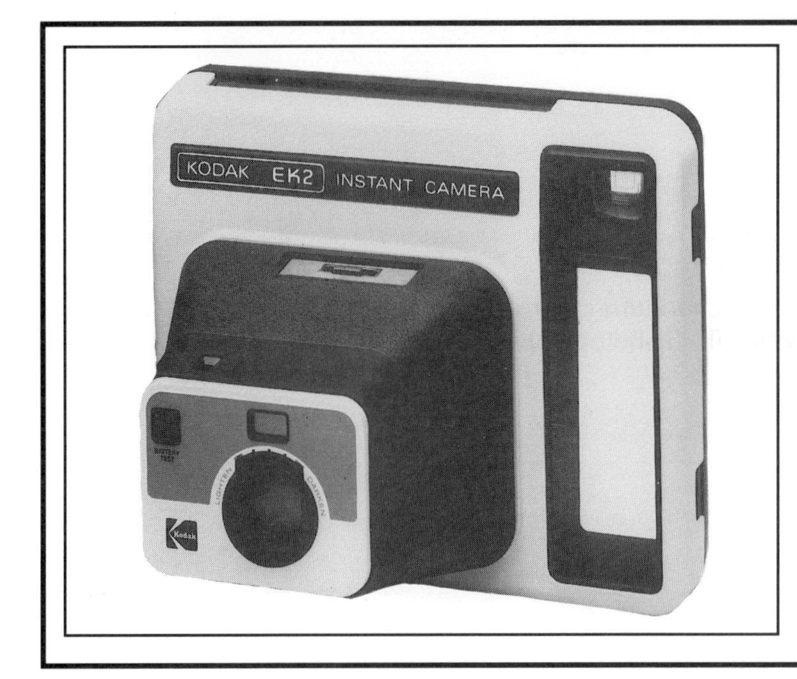

Kodak entered the instant picture market with a camera to rival Polaroid's SX-70 process. One of the Kodak cameras that started it all was the EK4.

and promotion industry, this was a 110 camera for sixteen exposures on colour negative film. With a single speed 1/80 sec shutter, fixed f/11 aperture and fixed focus lens, the camera was synchronised for use with Magicube flash and even had a double exposure prevention device.

After the film had been exposed, the whole camera was returned to Rankolor laboratories where it was broken open and the film processed. For the price of the processing, the user then received his sixteen colour prints, complete with a new camera ready loaded with a new film.

The following year, Kodak took their instant system a step forward with the introduction of the EK2 Instant Camera, a simple point and shoot model that left the photographer with little else to do but just that. After exposure, a handle on the EK2's side was turned to eject the developing print.

'The world's simplest camera'

But at that time, Polaroid were busy launching a new, simplified SX-70 camera that fell into much the same market as Kodak's EK2. They called theirs the Polaroid 1000 and advertised it as 'the world's simplest camera'. It was a claim that was well justified because they had improved even on Kodak's simplicity by having the print on theirs ejected automatically. The user quite literally had nothing to do but point the camera,

A decade before disposable cameras took off in a big way, Rank tried out the idea with a throwaway 110.

press the shutter release and wait for the developing colour print to be delivered into his hand a second later.

The end of the 1970s saw more and more automation being built into cameras. Electronic flashguns no bigger than an inch square began appearing as standard built-in features on many models such as the Vivitar 35EF released in 1977. This particular model was a full frame 35mm non-reflex camera with automatic exposure control as well as the built-in flash.

The viewfinder showed, besides the scene to be taken, a flash ready light, aperture scale, bright frame line, parallax correction mask and underexposure warning to tell the user when to switch on the flashgun. All this in a camera that measured only 5 1/2 x 3 x 2 in.

One company, Sunpak, famous up until then for their electronic flashguns, even boasted that they had fitted a camera in a flashgun rather than the other way around.

Holography

It was in the 1970s that stereo photography finally came into its own, though with a system that could never even have been dreamed of back in the days when Wheatstone first approached portrait photographers with his request for stereo pairs. The new system bore little or no resemblance to anything that had gone before. It was called holography and the three-dimensional picture that it produced was called a hologram.

Not that there was anything totally new about the system. A hungarian scientist named Professor Dennis Gabor first made a hologram in Britain in 1947, using mercury vapour lamps. The result was blurred and dim, but it was three-dimensional.

It was the introduction of lasers in the 1960s and their subsequent application to holography that finally made the hologram a viable proposition, both as an art form and in industry.

So what did a hologram of that time look like? It was a plate that could be viewed without the need for special glasses or any of the viewing apparatus that was needed before the stereo pictures. The viewer simply looked at the plate and saw a three dimensional picture apparently suspended in mid air.

By 1977, most photographers, amateur and professional alike, had been converted to automatic cameras, and the big discussion of the day was which type of automation was best. In its simplest terms, this could be broken down into two forms: shutter priority (you set the shutter, the camera sets the aperture) or the more popular aperture priority (you set the aperture, the camera sets the shutter speed).

The merits

Arguments raged over the merits and/or disadvantages of each system. But when it came down to it, most agreed there was something to be said for each. If you were photographing something that demanded a certain shutter speed, you needed shutter priority; if you were working on a subject where depth of field was important, then you needed aperture priority.

What everyone agreed on was that the ideal solution would be a camera that gave the choice of both. But such a thing didn't exist until late 1977 when Minolta launched the XD7.

Holograms in brief

To produce a 1970s hologram, a laser beam was first split into two separate beams, known as the reference beam and object beam. By means of a lens, the former beam was spread and directed straight onto the photographic plate's emulsion. The second beam was spread and reflected off the object and onto the plate. This formed an interference pattern on the emulsion where the two beams met.

The plate was then developed in a similar way to an ordinary photographic film or plate but instead of normal fixing, the unexposed halide was desensitized and the silver dissolved away by a special bleach. The result was a transparent plate that appeared to have nothing on it.

When a reconstruction beam was shone from a laser or other source, at the same angle as that of the reference beam at the time of taking the hologram, the three-dimensional image appeared.

At this point, the image appeared to be hovering behind the plate, but by making a second hologram from the original, a true-to-life stereo image of the object appeared in front of the plate.

The first multi-mode SLRs - the Minolta XD-7 (left) and the Canon A1.

The first camera in the world to offer a straight choice between aperture priority, shutter priority or complete manual override was the Minolta XD-7, released in the Autumn of 1977 and available to the general public by early 1978.

The Minolta XD-7 used a simple three-position switch to select whichever mode the photographer needed for the job in hand. All information such as shutter and aperture chosen was displayed in the viewfinder and focusing was helped by microprism and split-image aids.

Along with the XD-7, Minolta also released a motor wind and a remarkable new flashgun known as the Auto Electroflash 200X. The two went well together because the flashgun was capable of keeping pace with the motor wind's two-frames-per-second running time.

Canon's answer

It was followed, early in 1978, by Canon, who launched the A1, with not just one but five modes of automation. To Minolta's basic three, the Canon A1 added stopped down aperture priority and - for the first time on an SLR - programmed mode: a metering system that left the camera to select and set both aperture and shutter speed for correct exposure.

With this advance in metering design, soon to be copied by all the other SLR manufacturers, single lens reflex cameras became almost as easy to use as snapshot cameras, allowing users to merely point and shoot with a guarantee of correct exposure every time. Only one thing still stood in the way of a guaranteed perfect picture - the photographer still had to focus the lens. Automatic focusing for SLRs was still a few years off when Canon launched the A1, but automatic focusing arrived in compacts around this time with the launch of the Konica C35AF.

The autofocus age had arrived.

11. Automation all the way

THE SCENE IS a press conference, held in the UK at the end of 1977, for a camera due to be launched in the Spring of 1978. Konishiroku UK, the then importers and distributors of Konica cameras, had invited the press to see the world's first truly viable autofocus camera.

The idea wasn't new. There had, in the past, been a cine camera that had a built-in pendulum. Before beginning to shoot, the user pointed the camera at the base of the subject, the pendulum moved to remain perpendicular as the camera was tilted and, from that movement, the mechanism calculated and set the distance.

There was, at that time, evidence too of an autofocusing lens from Nikon that relied on part of the image being scanned by light-sensitive cells to find the maximum point of contrast - at which the image was reckoned to be at its sharpest, from which information the lens automatically focused. Unfortunately, the lens - an 80mm focal length - was nearly a foot long and weighed six pounds!

So it was a dubious collection of journalists who arrived for Konica's press conference - all of whom went away pleasantly surprised.

The camera being launched was the Konica C35AF, an automatic compact non-

Konica C35AF, the world's first truly viable autofocus camera, launched at the end of 1978 and in the shops by the following Spring.

reflex model with built-in electronic flash, a 38mm f/2.8 lens and programmed shutter speeded at 1/60, 1/25 and 1/250 sec. The auto-focusing mechanism worked by way of two windows. A computer control within the camera compared the images from these two viewpoints, used the information to calculate the distance and set the lens accordingly, together with the right exposure - all in the time it took the photographer to press the shutter release.

Such was the novelty of autofocus in those days, that Konica laid on an interesting demonstration to prove - short of taking pictures, developing them on the spot and showing them around - that it actually worked. A focusing chart was placed in the film plane of the camera and the whole thing was mounted in front of a projector so that the camera was now acting as the projection lens. A young Japanese lady then stood at the far end of the room, holding a white screen, on which the camera automatically focused the image of the focusing chart. She moved a few steps forward and the image naturally went out of focus. Then the camera's shutter release was pressed and the image snapped back into focus. It worked all the way down the length of the room. The photo press was convinced and went away muttering that this was the start of something big - a new age in photography.

A range of styles

During the following decade, camera design went through more innovative changes than it had in perhaps the whole of the past century. Autofocus was just one of the many innovations. Thanks largely to advances in microchip technology, taking new leaps every day, cameras, the way they were used and the people who used them were about to change for ever.

Camera styles of the time were many and varied. A glance at a photographic magazine from March 1978, dealing with how to choose a new camera, sees a wealth of different styles.

The 110 format was handled with not only examples of basic snapshot cameras, but also a 110 single lens reflex from Minolta. The Minolta 110 Zoom sported a 25-50mm f/4.5 macro-zoom lens and a shutter from 10 sec to 1/1000 sec, all in a camera weighing under 16 oz.

For 126, there were cameras like the Balda CLS from Germany featuring an f/5.6 three-element lens, an automatic electronic shutter speeded from 8 sec. to 1/1000 sec. and a brightline viewfinder with parallax correction and under-exposure warning. The camera was synchronised for flashcubes.

Then there was half-frame, epitomised by the Olympus Pen EE-2, which had auto exposure, but focusing by a scale of symbols.

The ever-popular 35mm format was covered by sophisticated-for-the-time single lens reflexes like the Canon F1 and Olympus OM2, as well as all the other big name camera makers. But this was also a time when the compact 35mm camera was coming seriously to the fore, a style which was soon to dominate the photographic market. In 1978, a camera like the Canon A35F best summed up the shape of the day's compact. The camera featured automatic exposure, a 40mm f/2.8 lens and focusing by means of a dual-image rangefinder in the viewfinder. Built-in flash, becoming increasingly important in this style of camera, popped up from the top plate at the press of a button.

Quality coupled rangefinder cameras like the Leica M4-2 were to be seen, but at that time, these were still being optimistically rivalled by cameras like the Kiev, which was a Russian-made direct copy of the pre-war Contax II.

Further up the film format scale, there were also a number of roll film reflexes on the market, using 120 film. They included cameras like the Mamiya 645 that took fifteen frames to a film, each 6 x 4.5 cm; the Rolleiflex SLX for twelve 6 x 6 cm pictures and the Mamiya RB67 Pro S that took ten 6 x 7 cm frames.

During the 1970s, the 6 x 4.5 cm format, taking fifteen exposures to a roll of 120 film, began to prove popular as a rival to 35mm for a few medium format camera manufacturers. The Mamiya M645 epitomises the style.

Ultrasonic autofocus

In 1978, Polaroid became the second camera manufacturer to come up with an autofocus system, though one that worked differently from Konica's. In the Polaroid system, an ultrasonic beam of sound was emitted from the camera and bounced off the subject. The time taken for it to be emitted and echo back gave the information needed to make the autofocus mechanism work. The camera chosen for the launch of the system was Polaroid's updated SX-70, thus Polaroid became the first manufacturer to produce an autofocus SLR.

Photokina, at the end of 1978, saw the debut of quite a few new camera designs. First came the Pentax Auto 110, the first true system SLR for 110 film. Minolta had tried a 110 SLR before, but it had never quite taken off, possibly due to its very unconventional design. The Pentax actually looked like an SLR in miniature.

The Pentax Auto 110

The Pentax Auto 110 was the smallest SLR around in the 1970s. It had an eye-level pentaprism viewfinder, centre-weighted TTL metering and a fully programmed exposure system that automatically selected both shutter speed and aperture in a range from 1/750 sec. at f/13.5 to a full 1 sec. at f/2.8.

Standard lens on the 110 format was a 24mm focal length and an 18mm wideangle plus 50mm tele were also sold. Close-up lenses and filters were available for all lenses. The camera also had its own flashgun and powerwind. The shutter mechanism was behind the lens, claimed as revolutionary by Pentax at the time, but really only a variation on two other cameras we have already seen - the Ensign Rollfilm Reflex back in 1922 and the Mecaflex in 1953.

The Pentax Auto 110 appeared in the shops early in 1979.

The second innovation at that Photokina came from Konica with the FS-1 - the first SLR with a built-in motor drive. Cameras were getting smaller all the time in the late 1970s and this year also saw the start of a very popular series from Olympus with the first Olympus XA. It took the form of a full-frame 35mm rangefinder camera, using aperture-priority automation in a neat and unique design that featured a dust cover that slid across the lens when not in use. The whole thing measured only 4 x 2$^1/_2$ x 1$^1/_2$ in.

That year's Photokina saw an ever-increasing number of compact 35mm cameras with built-in flash appearing and, the beginnings too of autofocus taking over in the style. The first 35mm compact after Konica with the feature was the Yashica Auto Focus, which was the first of its type to incorporate a focus lock device. Chinon and Cosina were also getting in on the act.

A new stereo system

The 1980s also saw the emergence of yet another stereo system, this time from a company called Nimslo. The first stirrings of what was to be a unique system came at this time with the launch by Nimslo of a device called a Computrack. It was really no more than a metal platform, on which any 35mm SLR could be mounted and slid along to predetermined places to take four pictures of any one subject. These four pictures were then optically combined at the printing stage and the result was a three-dimensional photograph that could be seen by the naked eye without any other accessory or viewing apparatus.

By the end of 1980, at that year's Photokina, the device had been turned into a four-lens camera no bigger than a standard 35mm compact. It took four pictures, each from a slightly different angle, simultaneously on 35mm film, and these had to be returned to Nimslo for processing. The special processing involved dividing each picture into strips, then printing them on special paper that incorporated minute raised ridges. When you looked at the picture, one eye saw one side of each ridge and the other eye saw the opposite side; thus slightly different views were seen with each eye and the 3-D effect was complete. The prints had to be held at a certain angle to see their effects to the full, but the fact is that the system worked. Nimslo was proclaimed as the next step in the evolution of the camera. The name, some said would stand beside the likes of Daguerre, Eastman and Land. They were wrong!

That same year saw the debut of other new cameras which had far more lasting effects. They included the Nikon F3, the company's first professional camera to rely more on electronics than mechanical systems; the launch of Pentax's first professional SLR - the Pentax LX; and a new 110 format SLR from Minolta that, unlike the company's first attempt at the style, looked like a small but conventional SLR; it sported a zoom lens as standard.

One camera, four lenses, three dimensions... stereo photography rises again, this time in the shape of the Nimslo.

The end of 1980 also saw an early attempt at an idea that had been in the air for some time - the autofocus 35mm single lens reflex. Early attempts at making the idea work came from Ricoh and Canon, each of whom produced, not an autofocus camera, but an independent autofocus lens for use with a conventional SLR. The Canon lens, naturally fitted only Canon cameras, but the Ricoh had a K-mount, making it suitable for a range of models including Ricoh and Pentax. Both lenses were extremely bulky, adding large rectangular units to the top of the barrel.

Autofocus technology

The Canon lens was a 35-70mm f/4 zoom which worked by a system known as 'Solid State Triangulation'. Inside the lens, hundreds of sensors were arranged in a linea array and received, side by side, two beams of light from the subject. One covered only the subject detail at which the lens was pointed, the other covered the entire scene. So the centre spot image of the actual subject fell onto different points in the array, according to the distance. By comparing the difference between the two sets of information, the lens calculated the distance and a micromotor rotated the lens to the correct position. In use, the photographer had only to point the camera at his subject, then press a button on the side of the lens to achieve accurate focus.

The Ricoh lens was a standard fixed focal length 50mm f/2 that focused automatically from 1 metre to infinity. Manual override was available for closer distances.

At this time, however, a serious autofocus 35mm single lens reflex was still a dream and any new product launch still involved manual focus in the SLR world. Canon, for instance, despite the fact that they now had an autofocus lens for their SLRs, still launched their next important landmark without any thought of autofocus features. It was an unusual step

Early attempts at SLR autofocus, courtesy of an autofocusing 50mm K-mount lens from Ricoh.

that updated an already popular camera without truly changing its name. So the Canon F-1 became the Canon new F-1 or F-1n.

The new camera was rather old fashioned by the current day standards, but one which nonetheless appealed at once to professionals many of whom, during the past few years had begun to become disillusioned with Nikon, whose electronic F3 wasn't proving quite as reliable as the old mechanical F2. The Canon, then, won many friends by producing a professional SLR with an electronic shutter which, when the battery was removed, could be used mechanically from 1/90 - 1/2000 sec. The camera was a manual match-needle meter model - again, a little old fashioned in a world in which most SLR viewfinders (including Canon's own A1) were flashing lights at the photographer rather than moving needles.

Fitting a different finder to the camera added aperture priority automation and adding a motor drive gave shutter priority. The camera also offered a range of 13 different focusing screens in three groups that added up to 32 different focusing-metering combinations.

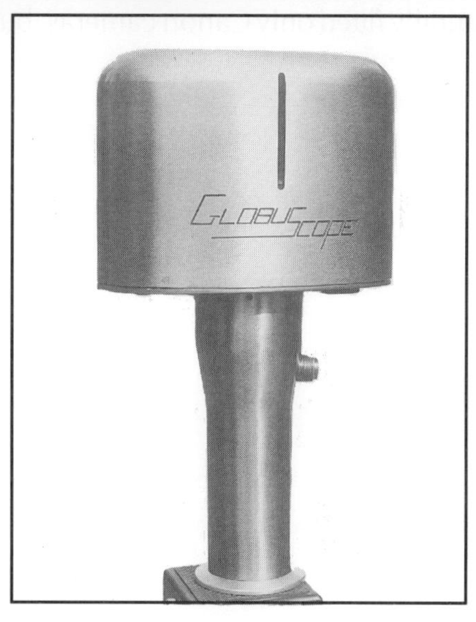

The highly unusual Globuscope. Above, the camera ready to shoot; below, how the film travelled inside the camera; top, the format it produced.

Despite the fact that the new F-1 was a little old-fashioned, even at its launch date, it quickly won praise from professionals who wanted a basically sound, rugged camera and remained popular for another next decade, when it was gradually phased out in favour of autofocus models, including a professional version.

During the history of the camera, we have seen many so-called new styles, based on old ideas and one which re-occurred at regular intervals was the panoramic camera. The 1981 version was called the Globuscope, an unusual design that worked in a similar fashion to the old Kodak Cirkut in which the camera revolved on its tripod, while film moved past a shutter slit in the opposite direction.

The Globuscope took standard 35mm film. The 25mm lens offered apertures between f/3.5 and f/22, while the shutter 'speed' was controlled by the width of the slit in a piece of metal that was screwed into the film plane before the film was loaded. Revolving on a sturdy handle that incorporated a clockwork motor for the purpose, the Globuscope produced an image wider than four standard 35mm frames put together - 160mm wide by the standard 24mm deep.

Kodak's new format

Around this time rumours were rife that Kodak were about to do it again - launch a new film format, that is. The new format, the rumours went, would involve discs - something which at that time was synonymous with computers and video. A number of different designs were speculated about, but such was the secrecy of the launch that no one really knew what the cameras would like before the official press launch - a rare thing in the photo industry. In fact, not even invitations to the launch mentioned the emotive word 'disc', though everyone suspected that was why they were being wined and dined, in the UK, at six o'clock in the evening. The reason was the new disc system was being timed to be announced at exactly the same time across the world, and what was noon in New York was six in the evening in London.

The new disc cameras, in the end, had nothing to do with video or computer discs. The discs in question were actually flat circles of film that revolved in a neat little flat camera to give 15 exposures to a loading. The system was launched with four cameras of varying degrees of sophistication - the Kodak 2000, 4000, 6000 and 8000. Each camera featured a 25mm standard lens, producing images just 10 x 8mm - rather grainy on the ISO 200 speeded disc film, even on conventional small-size enprints.

The flagship of the Kodak launch was the Disc 4000. It featured a fixed focus 25mm focal length lens with a maximum aperture of f/2.8, automatic exposure, automatic motor drive between exposures and an in-built flash that turned itself on automatically as light levels dropped. The camera measured 4 3/4 x 3 x 3/4 in.The style eventually failed because of image grain from the small film size against the quality being seen in the fast-developing 35mm compact market.

Disc mania hit camera manufacturers in the mid-1980s, following Kodak's launch of the format in 1982. It wasn't long before every manufacturer started making their own disc cameras - and not long after that they stopped. This one is the first Kodak disc.

The autofocus SLR came a little closer in 1982 with a Canon idea called focus aid. It was to be found on their new AL-1 single lens reflex whose viewfinder showed, not the usual rangefinder for accurate focusing, but two red LED-lit arrows. Turning the lens's focusing ring in the direction of the lit arrow brought the subject into focus, at which point the red arrow as replaced by a green light.

The first still video

Another innovation sneaked quietly in during 1982, one which didn't have much impact at the time, but one which nevertheless was due to be an important influence on the shape of cameras to come. This was the Sony Mavica - the name a corruption of *Magnetic Video Camera.* The camera was an SLR with standard, wideangle and tele lenses. The big difference between this camera and any that had gone before was that it didn't use film. Instead, it recorded its images digitally on a small computer disc, called a Mavipack, 50 exposures to a loading. This, in turn, could be used to play back the pictures on the camera's own viewer or a standard television, via the camera's own playback machine. Hard copy prints could also be made with the Mavipack copier.

The Sony Mavica was announced to the press, but never really went into production. Even at Photokina that year, it was difficult to find and was eventually

The Sony Mavica - first still video camera, but not one that eventually made it to the consumer market.

traced to the electronics section of the show, rather than the photographic areas.

Photokina, that year, had little time for what was to become still video cameras. This, undoubtedly, as far as that show was concerned, was the year of the disc.

Kodak, in the past, had always done well launching new film formats with their own cameras, then making a lot of sales on the new film size as all the other manufacturers leapt on the bandwagon to produce their own versions of the cameras The launch of the disc earlier in the year was no different and, by the time Photokina came around in the Autumn of 1982, everyone and anyone had a disc camera to show off. Rivalling Kodak were cameras from Halina, Hanimex, Keystone, Nova and Fuji, to be joined a little later by Minolta and Konica, to name but a few, each with a different model in varying degrees of sophistication.

Instant 35mm

At the same show, Polaroid announced their 35mm Instant Slide System with three 35mm films, one for colour, one for continuous tone black and white and the third for high-contrast black and white, each of which could be shot in any standard 35mm camera, then processed within a few minutes without the need for a darkroom.

The Nimslo saga was still going on at that year's Photokina. First seen more than two years before, it still hadn't been officially launched but promises of a 1983 Spring date were announced at this time. The Nimslo process was, in all truth, a remarkable one, but the hard fact is that, when it was eventually launched, it just never caught on with the public. Snapshotters never got the hang of it, more dedicated amateurs thought it a gimmick and professionals saw no real need for it. Within a few short years, the Nimslo system seemed to be dead and gone. It wasn't, however, forgotten, as we shall see when we get to the end of this chapter.

All of which shouldn't detract from the way things were going with two of the most popular styles of camera: the 35 single lens reflex and compact.

SLRs were undergoing almost daily changes with a whole new breed beginning to emerge. It was epitomised by the first of the style in the Canon T50, a programmed-automation SLR with built-in motor drive that gave snapshotters and photographers with little or no real knowledge of basic techniques the chance to take perfectly-exposed SLR pictures. The camera, in a new sleek design, also incorporated a built-in motor drive. It was followed by the Canon T70 that added extra modes to the same basic style and it was this camera, perhaps more than any other that set the style for similar models from most of Canon's opposition over the next few years.

Compact revolution

Compacts, on the other hand, were gaining more ground everyday. As the 1980s moved along, most manufacturers incorporated auto exposure, autofocus, automatic film load, wind and rewind, and built-in flash. Perhaps the most important of these functions as far as the marketplace was concerned, was the automatic film loading. If there was one thing that had, up until then, separated the snapshotter from the serious amateur photographer it was the difficulty the former found in loading a cassette of 35mm film. That's why 126, 110 and then disc cameras were so popular.

But with automatic loading handed to the market, even the rawest beginner to photography could start with 35mm, a format which was bound to give better all-round quality than 126, 110 or disc. What's more, the auto exposure and autofocus features ensured technically perfect pictures with almost every press of the shutter button.

Suddenly, 35mm became the format, not only for professionals and serious amateurs, but weekend and holiday snapshotters as well. Almost overnight, 126 and

Polaroid's instant slides

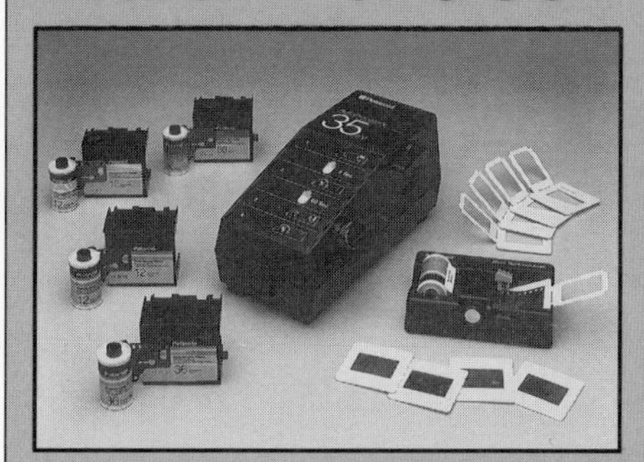

In the Polaroid 35mm Instant Slide System the exposed film was placed in a desk-top processor, along with a special processing pack that came with each film.

The lid of the processor was closed, a lever pushed down, a handle turned and after, a minute's wait, the lever returned to its original position and the handle turned again. During this procedure, processing fluid from the special pack was coated onto a strip sheet which was then laminated to the exposed film.

When the processor was opened, the strip of slide film, in colour or black and white, was processed, dry, ready for mounting and projecting.

110 began to die and it wasn't long before they took the disc cameras with them.

Every year through the 1980s took the camera industry closer to a true autofocus SLR and in 1985, another major milestone was reached, once again with a camera from Canon. It was the T80, very much like the T50 and especially the T70 in styling, but with a 50mm autofocus lens as standard and the option of 35-70mm or 75-200mm autofocus zooms as well. The Canon T80 went a long way towards addressing the idea of an autofocus single lens reflex, but it was an ugly camera, made too bulky by the size of the autofocus optics, a fact which Canon obviously recognised, since their next launch was the T90, a manual focus camera, but one which took computerised electronics about as far as any camera of the time could go. It even featured an interface with Canon's own computer range.

It in fact fell to Minolta, later the same year, to come up with what was the first truly viable autofocus 35mm SLR with the launch of the Minolta 7000. The camera represented the peak of SLR technology in 1985 and its design illustrates well the way a whole new generation of 35mm single lens reflexes from all the top manufacturers had developed over the previous few years - and the way they were destined to go in the future.

First dual-lens compact

By now compact cameras were really beginning to dominate the market, bringing ease of use with guaranteed results to all. Most, by this time, featured autofocus, auto exposure and automatic film loading, to which Minolta - again - brought a new feature. The Minolta AF-T was the world's first 35mm compact camera with a choice of lenses. It had a 38mm lens, the standard focal length on compacts of the time, and a 60mm that could be brought in at the flick of a switch. Minolta called this a tele lens, even though its focal length was actually pretty close to the standard 50mm on SLR 35mm formats. Even so, it was a new landmark and one which was beaten the very next year, in 1986, by Pentax with their Zoom 70 compact which, as its name, implies, featured a zoom lens with a maximum long focal length of 70mm.

But it was Minolta who still held sway with camera landmarks of the time. They followed the 7000 with the 9000, a new version for the professional market for professionals and, in 1988, by which time most major manufacturers, including Canon, Nikon, Pentax and Olympus had autofocus SLRs of their own, Minolta came up with another first. Their Dynax range of autofocus SLRs featured small program cards that could be inserted in a slot to program the camera's computer to give it new and extra functions. A Minolta Dynax camera could be all things to all photographers, according to the way the photographer programmed it.

This was also the year that Kodak bowed out of the instant photography market. After twelve years of court hearings, Polaroid won the day on their patent battle and

The Minolta 7000 autofocus SLR

The Minolta 7000 was an autofocus 35mm single lens reflex that did away with most of the controls traditionally associated with SLRs.

The top plate featured a Liquid Crystal Display, bringing information on shutter speeds, apertures, film speeds, exposure modes, etc., all of which were brought into play by a few simple push-button controls.

To change exposure modes, the photographer held down a 'mode' key, while pressing an 'up' or 'down' button that took the camera sequentially through the options. Those options included program, aperture priority, shutter priority and manual.

The big difference between the Minolta and its then current rivals, however, was its body-integral autofocus system which meant. for the first time, an autofocus SLR with a fairly normal-sized lens. First pressure on the camera's shutter release activated the system and the lens focused on any object found central in the viewfinder. A focus lock device naturally allowed more adventurous composition.

The Minolta 7000 also featured automatic film load, wind and rewind, as well as automatic flash exposure.

Kodak were forced to withdraw all instant photography products which, by that time, included not only cameras but also professional audio visual copying systems and equipment for both prints and instant colour slide film.

In 1988, the throwaway camera idea again surfaced, with a little more sophistication than before. These cameras were basically cassettes of 35mm film with simple, fixed focus, one aperture/shutter speed cameras built around them. The Fuji Quicksnap was first, quickly followed by the Kodak Fling, each selling for little more than the cost of a cassette of 35mm colour print film. The same year, Fuji followed up with the Quicksnap Flash - same idea but with an in-built flashgun, followed in 1989 by the Kodak Stretch, a panoramic version, plus the Kodak Weekend, which was weatherproof.

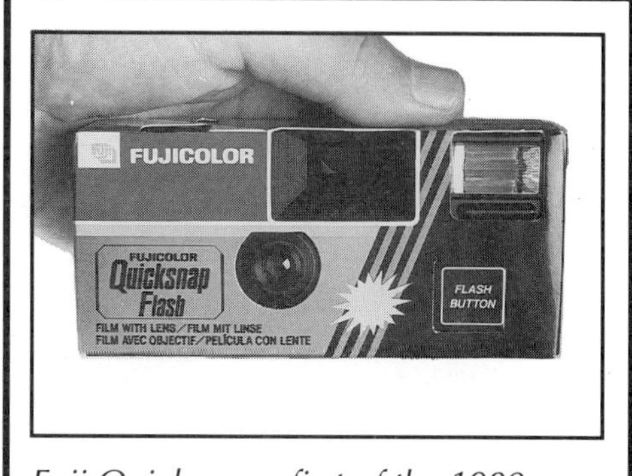

Fuji Quicksnap - first of the 1980s style of throwaway camera.

The next major landmark in the history of the camera came at the end of 1988 with the launch of the Olympus AZ-300, a non-reflex camera, but one which couldn't exactly be described as a compact. The AZ-300 was a complete new style of camera, that offered many of the advanced features of an SLR, but with the ease of use of a compact. Its very different design offered a fixed 35-105mm zoom lens, programmed exposure using a shutter speeded from 1-1/500 sec., through-the-lens metering, built-in film wind and a built-in flash that featured an optional function that claimed to defeat red-eye.

A new name

At a loss for something to call this new breed, they became known as bridge, hybrid and even zoom cameras. Needless to say, the Olympus was soon followed by others of a similar ilk, but each with its own little extra. The Chinon Genesis offered a similar basic idea in an even stranger shape and finished it all off by making it a half-frame camera to take seventy-two 24 x 18 mm frames on a standard cassette of 35mm film. The same year saw the arrival too of the Ricoh Mirai, which took the same idea again and turned it into an SLR

The arrival of these hybrid cameras, in truth, said a lot for the state of the camera industry at the end of the 1980s. It was a time when it seemed that cameras had gone, if not as far as it was *possible* for them to go, then at least as far as it was remotely *necessary* for them to go. It was a time when SLRs incorporated every feature that was needed and a lot more than would never be used, while compacts, dominating the

The Ricoh Mirai hybrid SLR

The Ricoh Mirai, launched at Photokina in 1988, was the first of the Hybrid style of cameras to be released as a single lens reflex.

Taking on an entirely different shape from any conventional 35mm SLR seen before, the Mirai had a fixed standard zoom lens with a focal length range from 35-135mm. Maximum aperture was f/4.2 at the wide end of the focal length range, f/5.6 at the long end; shutter speeds were from 32 full seconds through to 1/2000 sec.

Focus was automatic and exposure was programmed, based on a centre-weighted through-the-lens metering system with automatic backlight control. A built-in flash was synchronised at 1/100 sec and its output was automated by through the lens lighting reading. Macro focusing was possible and the camera had built-in film wind.

The reflex viewfinder had indicators for focus confirmation, shutter speed and aperture in use, flash ready, exposure lock and macro details.

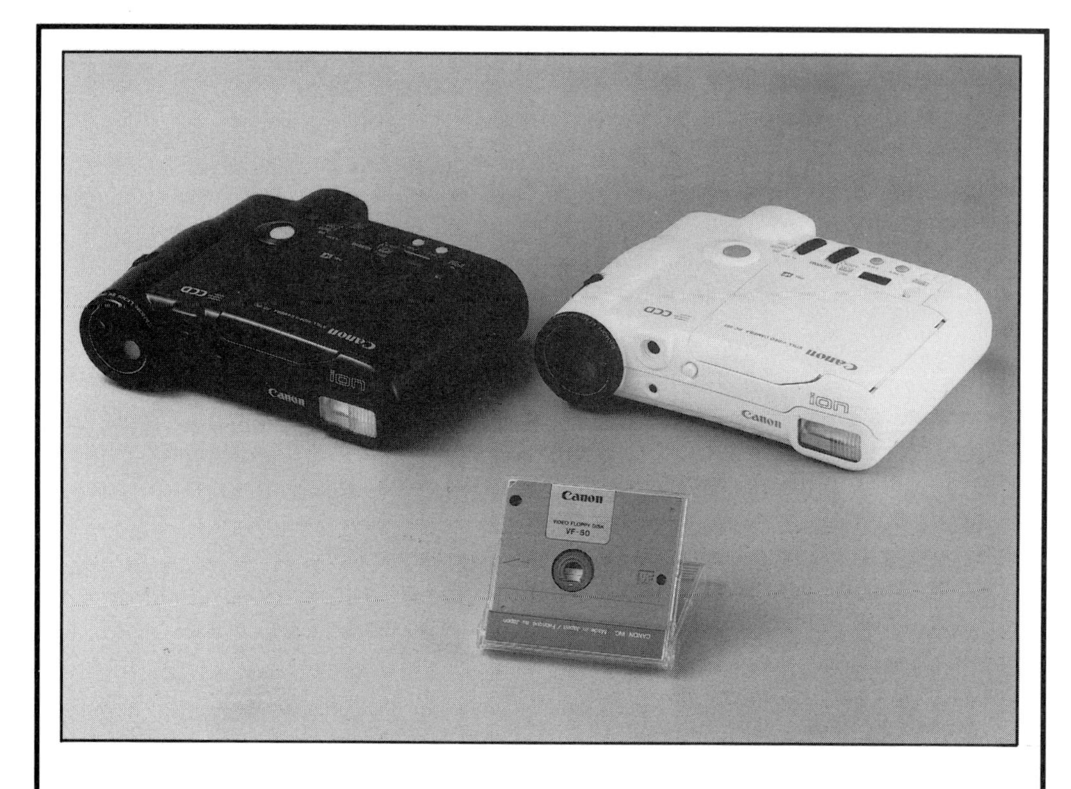

The Canon ION - available in a choice of colours - together with the small disc that replaced the need for conventional film. Right at the end of the 1980s, the new style marked the start of the next era in the history of the camera.

market as far as sales were concerned, were just getting too clever for the market they were aimed at. It seemed that never a week went by without every major camera manufacturer launching a new compact - as many as 60 new models appearing over a two-month period at one time - and each one offering very little new over its predecessors, other than a new name, a different number on the top plate and some very minor feature. Manufacturers were even reduced to introducing messages saying *Happy Christmas* that could be printed on the film as the picture was taken.

Then, as the decade closed there was a surprise for stereo fans. The Nimslo system, thought by most to be dead for ever, suddenly reared its head again, this time from Hong Kong. The move came about when the Nissei company bought the rights to the Nimslo lenticular 3-D system for their own use, then launched their own version of the old Nimslo camera, calling it the Nichika N8000.

The four-lens camera had a fixed shutter speed of 1/60 sec, each lens offering a choice of three apertures from f/8 to f/19, corresponding to sunny, partly sunny and cloudy conditions. Strangely, although the camera had less features than the old Nimslo, it was nearly twice the size.

The real truth of the matter at this time though, was that by the time the decade drew to its end, there were no new features that were really needed on cameras and those that were introduced seemed to be there more for marketing purposes than for any genuine usefulness. It seemed that conventional camera design really had gone as far as it could go. Then Canon launched the ION.

A new era...

First seen at the end of 1989, the Canon ION (short for Image Online Network) was the first truly viable still video camera for the consumer market. As has already been noted, it was by no means the first. The Mavica had been seen seven years before, many manufacturers, including Canon and Konica had shown examples at Photokina every year since and Canon had even had a test run for professionals at the 1988 Olympics. But this was the first time a still video camera had arrived in the shops, aimed straight at the consumer, at a price comparable with a mid-range SLR.

The Canon ION used a magnetic floppy disc instead of film to capture its images. They were then played back, immediately if desired, by connecting the camera by cable directly with any standard television. Up to 50 full-colour pictures could be captured on a 54 x 60 mm size disc, each one capable of being erased and re-shot at the touch of a button. In shape, the ION looked like a pair of binoculars. All functions were automatic and the lens was fixed focus type. Self-timer and flash were built in and the camera was capable of taking single shots or three images per second.

So, as perhaps one of the most innovative decades in the history of the camera drew to a close, a new breed of camera had arrived. It would be a long time before still video discs totally replaced film and more traditional styles would be around and launched for a long time to come. But, without any doubt at all, the future had arrived. It was, in many ways, the end of an era.

And the start of a whole new age in the history of the camera...

Index

A

B

C

D

E

F

H

I

J

K

L

M

N

P

R

S

T

U

V

W

Y

Z